The Mathematical Traveler

Exploring the
Grand History of Numbers

The Mathematical Traveler

Exploring the Grand History of Numbers

CALVIN C. CLAWSON

PERSEUS
PUBLISHING

A Member of the Perseus Books Group

Copyright © 1994 by Calvin C. Clawson

Library of Congress Control Number: 2003102872
ISBN 0–7382–0835–3

Perseus Publishing is a member of the Perseus Books Group.
Find us on the World Wide Web at http://www.perseuspublishing.com.

Perseus Publishing books are available at special discounts for bulk purchases in the U.S. by corporations, institutions, and other organizations. For more information, please contact the Special Markets Department at the Perseus Books Group, 11 Cambridge Center, Cambridge, MA 02142, or call (800) 255–1514 or (617) 252–5298, or e-mail j.mccrary@perseusbooks.com.

First paperback printing, April 2003

1 2 3 4 5 6 7 8 9 10—06 05 04 03

To my wife, Susan

Acknowledgments

Many people helped to make this book a reality. Special thanks go to my workshop friends who patiently reviewed the manuscript and offered many excellent suggestions: Marie Edwards, Bruce Taylor, Linda Shephard, Phyllis Lambert, and Brian Herbert. I also want to thank my professors at the University of Utah, in both the Departments of Philosophy and Mathematics for their care and dedication in teaching me to love mathematics.

Contents

ix

Introduction

This book is about numbers. However, it is not a book that addresses only formal mathematics. While it is true that mathematics is the study of numbers, the use of numbers is so interwoven with the fabric of human experience that to talk of all aspects of numbers is to delve into many of the very mysteries of human nature. While this book contains mathematics, it also invokes anthropology, biology, psychology, anatomy, history, and philosophy.

Human beings have been called "the tool-using apes," "the fire-using apes," and "the talking apes." However, it would be just as proper to call us "the counting apes." Numbers and counting permeate our culture. Try to imagine, if you will, our society without any numbers or counting. We are so attuned to numbers that we use them to measure every aspect of our lives. We mark where we live with street numbers. We call each other on the telephone with numbers. Our money is based on numbers, along with our calendars and clocks. When I woke up this morning, I turned over and looked at my digital clock. It was 6:45. I remembered today was the 17th, and that reminded me I had to pay a $110 car payment. While I ate breakfast, I listened to 97.3 on the radio: the stock market was up 14 points. Thus my day (and everyone else's) continued. We rely on numbers in order to function in our modern society.

The economy, our technology, and science all depend on our use of numbers, and this dependency is reflected in how early we introduce our children to numbers; one of the first things we teach them is how to count to ten. Yet even with this early introduction, many millions of Americans shudder when someone mentions mathematics—the study of numbers. "It's too dry," they complain. "It's too hard—all those symbols." But some of these same individuals go to Las Vegas and sit up all night to play a number adding game—blackjack. Or they play a dice number game— craps. Maybe they just stay home and play Monopoly, a game that requires them to add together the dots on two dice and then count how many places to move. If they land on someone else's property, they must count out the rent. If the property is unowned, they count out enough to buy it. The entire game is adding, subtracting, and counting with numbers. People love numbers. Sometimes they claim a disinterest in formal mathematics, but in actuality, whether it's playing a game, anticipating a check from the government, or "playing the stock market," people love numbers.

Given our secret delight in numbers, you will very likely enjoy this book. We begin our adventure by looking at the natural numbers—those numbers we first learned to count, namely one, two, three, four, and so on. By "so on," we mean we can count on forever and never reach the last number.

The natural numbers were the first numbers discovered by human beings. The first thing they did with them was count. In Chapter 1 we will define counting and then investigate that part of our brain responsible for counting. Next we will consider how old counting and numbers might be and how they were first used (Chapter 2). In Chapter 3 we will consider whether other species can count. In Chapter 4 we will look at numbers as used by the first farmers and then the first written records of numbers from the Mesopotamians and Egyptians. We will then review the early contributions to numbers by the Chinese and Native Americans (Chapter 5). Next we will learn of a great mystery about numbers that was first encountered by the ancient Greeks but not solved for over two thousand years—a mystery that eventually led to the strange irrational numbers (Chapter 6). In Chapter 7 we will introduce the negative numbers and our modern Hindu–Arabic number system, after which we will be ready to look at the concept of the infinite and how this idea impacted our beliefs about numbers (Chapter 8). By now we will have followed the evolution of numbers sufficiently to solve the mystery of the irrational numbers

(Chapter 9). Then we will free our imaginations and discover the strange transcendental numbers, such as π—numbers so peculiar that we cannot even write them down (Chapter 10). Next, we will investigate an entirely new class of numbers, the complex numbers, used extensively in science and engineering (Chapter 11). This will be followed by our last class of numbers, the transfinite numbers (Chapter 12). In Chapter 13 we will pause to consider some of the great number-crunching prodigies. At this point we will be ready to consider the philosophical side of numbers (Chapter 14). The last chapter will end our journey with a review of twentieth-century contributions and speculations about where mathematics is going in the twenty-first century (Chapter 15).

Because the natural numbers and our culture are so intertwined, we will actually see a microcosm of the history of mankind as we study them. It is a history filled with surprise, wonder, and delight.

Even though most of us have been rather cavalier about the importance of numbers, through the ages some have recognized the role numbers play. Over four hundred years before Christ, a Greek by the name of Philolaus of Tarentum said:

> Actually, everything that can be known has a Number; for it is impossible to grasp anything with the mind or to recognize it without this [Number].[1]

The great Greek philosopher, Plato, taught that numbers not only occupy a central role in our world but even lead us to truth itself. In a dialogue between his mentor, Socrates, and a friend, Glaucon, Plato had the Athenians comment on numbers.

> Socrates: And all arithmetic and calculation have to do with number?
> Glaucon: Yes.
> Socrates: And they appear to lead the mind towards truth?
> Glaucon: Yes, in a very remarkable manner.[2]

So what is the thesis of this book? What do we want to come to understand about ourselves by reading this volume about numbers? This book demonstrates that numbers, and therefore mathematics, are integral to our very nature, that we cannot function as human beings without numbers.

The counterthesis is simple and, unfortunately, one held by many Americans: While most of us count, and occasionally add or subtract when absolutely necessary, mathematics in general is an esoteric endeavor performed by old bearded men in the ivory towers of universities. It is not really important to our lives in the way that love, our families, or our careers are. Therefore, it is more of a bother than a help.

This volume will demonstrate the opposite. It will show that mathematics is buried deep in the very fabric of who we are. It is an activity that helps define us. If the day arrives that we travel to a distant solar system and visit other intelligent beings, they may ask "Who are you?" We will say with pride, "We are the counters; we know numbers."

CHAPTER 1

How Do We Count?

How deeply are counting and numbers buried into the very nature of being human? Is counting a minor skill we acquire after our brains are fully developed? Is it entirely dependent on cultural learning? Or, is counting somehow "hard-wired" into our brains? Is counting so basic to human beings that it is a part of our genetic inheritance? In other words, how do we count?

VOCABULARY

Before we begin our grand adventure, let's take a moment to discuss vocabulary. Whenever we want to communicate in a more precise manner, it helps to define new terms or redefine old terms in more exact ways. This improves understanding between people, whether they are engaged in a special skill, technology, or science. Yet, a specialized vocabulary also tends to shut out people who might be interested in learning a new technology or science. Some practitioners, such as doctors, psychiatrists, and lawyers, have been accused of manufacturing an entirely new language just to keep nosy outsiders at bay. We will not argue this point.

Mathematics does contain numerous special definitions to increase the precision of communication. Our approach will be to provide those new terms when needed, but we will always lean toward the simple statement rather than the complex or convoluted. Frequently, the new concept is not difficult to learn, only the strange new vocabulary is. Our goal is to increase our understanding of numbers, not just to learn new words. When we introduce a mathematical term, we will give it a short, straightforward definition. We will not introduce new words where they are not needed.

THE NATURAL NUMBERS

In order to get a sound understanding of just what numbers are, we will first investigate the natural numbers. Natural numbers are those numbers we are most familiar with; they are the numbers with which we count: one, two, three, four, ... The ellipsis points mean that we can continue to count as high as we want; there is no limit to our counting imposed by the numbers themselves. Of course, a practical limit exists, for we can count only as high as time allows. Sometimes the natural numbers are called "whole numbers" or "positive integers," and while these three expressions are equivalent, we will generally prefer to use "natural numbers."

There are three basic uses for numbers. When we want to know how many things are in a collection of items, we are after the *cardinal* number. The collection is called a *set* by mathematicians. Therefore, if we have a basket with eleven apples, the cardinal number of apples is eleven, and the set is the collection of eleven apples. A single item in a set is called an *element*. Hence, each apple is an element of the set of apples.

Cardinal Number: A number specifying how many items are
    in a set.
Set: A collection of items.
Element: One particular item in a set.

You can easily see how useful the idea of a cardinal number is to human beings. It answers the questions: How many cars are in the parking lot? How many dollar bills do I have in my wallet? How many children do my cousin, Wilford, and his wife Mavis have? The answers to all these questions are cardinal numbers.

Now we can define a second use of numbers. It is to show the relative order of things. When you go into a coffee store to buy your favorite bohemian java, you find a long line. How long will you wait to be served? You pull the number 47, noting that the sign above says, "Now serving number 35." Here, the number 47 does not tell you how many things are in a set, but where you are relative to someone else. There are twelve customers before you. Thus, the 47 shows your order in a sequence of numbers. This kind of number is an *ordinal* number.

We use many ordinal numbers. Your street address is an ordinal number. It does not count houses or people or any set of elements. It indicates where your house is in relation to other houses on your block and in the city. Anytime you have a number showing the relative position of things, you have an ordinal number.

Ordinal Number: A number that specifies the relative position of an element in a set.

The third use of numbers is simply identification. These are *tag* numbers. They do not count elements in a set nor do they show any relative order. Your phone number is a tag number. It does not identify how many phone numbers are in the set of phone numbers nor does it show relative position of your phone number to other phone numbers. Social security numbers are tag numbers too. Most bus and airline numbers are tag numbers. However, we are not really interested in tag numbers, because such numbers simply stand in place of names. In each case a name could be used instead of the number. Therefore, in the pages that follow we will be concerned with ordinal numbers and cardinal numbers, but not tag numbers. The natural numbers can be used for both cardinal and ordinal numbers. They can be used as cardinal numbers because they can specify the number of elements in a set. They can be used for ordinal numbers because every natural number is assigned a unique position within the sequence of numbers. The number five is always after four and before six. Hence, the natural numbers are ordered.

THE SPECIFICS OF COUNTING

From early childhood we are taught to use natural numbers. Our parents taught us to count to ten. Frequently this is done by pointing to successive fingers: one, two, three, ... As each number was named, our parents touched another finger. We thought it was all great fun and we quickly learned to mimic them. Soon we could name the number words in the correct order: one, two, three, four, ... Yet, just learning to say the correct words in the correct order is not counting. Counting requires something more: It requires that we answer the question, how many? Our parents helped us with this too. Your mother put three of your toy blocks before you. "How many, sweetheart? How many blocks?" She was trying to get you to count them. At first you may have just scattered them about.

But eventually you learned. You could point to things and repeat the correct names as you counted. You learned that the last name you said was the "number" your mother wanted. Did you understand the idea of "how many?" If you were very young, probably not. Yet, sometime before you began school, you likely did come to understand the "many-ness" of sets. Especially when your big sister said she didn't take a toy soldier, and you knew one was missing. You had counted them.

Here, we must stop and provide an additional word for counting. You probably learned counting as an activity. You pointed to successive fingers and said the correct number words. Now we will introduce a term dear to mathematicians—*mapping*. When you count, you assign one finger to exactly one word. You do not assign two fingers to one word nor do you name two words as you touch one finger. It is a one-to-one process. Each finger gets its own number word, but only one. Mathematicians call this a one-to-one mapping. It is at the very heart of our procedure for counting.

> *One-to-one mapping*: To assign exactly one and only one
> element from a set (e.g., number words) to each element
> in a second set (e.g., fingers).

At that point in your early childhood, probably somewhere between the ages of two and five, you achieved a great sophistication in thinking, for you learned two of the most difficult concepts in mathematics you will ever learn. You learned to conceive of things in collections or sets, and to understand that each set has a manyness which is identified by its cardinal number. You knew that different sets can have the same cardinal number or different cardinal numbers. You learned a procedure for counting and thus for discovering the correct cardinal number for a set. This is an astounding collection of skills.

We are now ready to ask the basic question: How do we count? What thought processes are used? What part of the brain is activated? Is there a specific area employed for this activity? Is it purely a learned skill or do we inherit it? Is language necessary to count, since we use words when counting?

The complete act of counting depends on three activities. First, we want to answer the question: How many? We have identified a set and wish to know the manyness of that set. The next two activities occur simultaneously. In sequence, we touch or point to each element of the set with our finger and say the appropriate number word. Pointing and saying the correct number word is, of course, our mapping activity. When we are finished, the last number word spoken is the cardinal number we are after, and our counting is complete. Hence, we have a need (1) to know how many, (2) to sequentially touch or point, and (3) to sequentially speak the number words. Thus, it would appear that those parts of the brain used in counting include the areas responsible for abstraction, sequential motor skills, and language skills.

Sometimes we do not actually point with our finger but only move our eyes from one object to the next. Yet, even moving the eyes is a sequential motor activity. In this case we have simply discontinued the pointing or touching, and continued with the eye movement. On some occasions we do not speak the number words aloud but only think of them.

Have we identified the smallest activities necessary for counting? If this were so, we would assume that counting is dependent on language, since we would have to learn number words before we could count anything. Hence, counting would be a skill acquired after (or no sooner than) language.

Does a form of counting exist that is independent of language? If so, this form of counting might be even older than modern rapid speech, and might not depend on those brain areas associated with language.

The book *Number Words and Number Symbols: A Cultural History of Numbers* by Karl Menninger is a classic contribution to the understanding of early counting and numbers.[1] In his work, Menninger points out that nonlanguage counting is not only possible but almost certain for early humans. Even up to the twentieth century, some primitive tribes had only two number words: one and two.* Everything else was simply "many." Yet, individuals from these same tribes were often required to keep track of the number of elements in a set of possessions. The words "one" and "two" were simply not adequate to account for thirty head of cattle, or eleven dogs, or sixteen baskets of grain. Accounting for, and keeping track of, a number of possessions, especially when such possessions included food, shelter, or safety from one's enemies, created strong incentives for people to keep track of the cardinal number of sets. If a dishonest neighbor keeps stealing your cows at night, and you remain ignorant of this fact, then you place yourself and your family in danger of starving.

Menninger points out that some primitive tribes, such as the Wedda tribe on the island of Sri Lanka, do not count by mapping number words to elements, but by mapping sticks or other items to the elements to be counted. Hence, to count a set of coconuts, a Wedda tribesman can assign one stick to each coconut. When finished, the bundle of sticks he holds in his hand is the "number" of coconuts. He cannot tell you how many coconuts he has, for he does not know the proper number word for the bundle's cardinal number, but he can hold out the bundle of sticks to "show" you that number.

At this point you may object that the Wedda tribesman is not really counting the coconuts. You may want to maintain that true counting is understanding the number sequence (one, two, three, four, . . .) and how to count with it. However, it doesn't matter to the Wedda tribesman whether we consider his activity as true counting or only as some sort of "stick-counting." What matters to him is that he arrives at the correct cardinal number for the set of coconuts by mapping sticks onto the coconuts and, hence, answers his question "how many?" By answering this question he is able to account for his coconuts, cattle, and other possessions. He does it through a manipulative skill that does not require the use of language.

---

*Here I use the word "primitive" to describe the condition of the tribe's culture and not to describe the physical attributes of its members. All human beings alive today are modern *Homo sapiens sapiens*.

Just what skills are necessary for his simple stick-counting? First, the tribesman must have the abstract idea of the manyness of a set. That is, he sees a collection of different items (individual coconuts) and realizes that they all share a common nature. He recognizes them as a set of objects and wants to account for their multiplicity. Second, he must plan to make a series of manipulative acts to map each stick to each coconut. This sequence of steps only makes sense when carried out to completion, since any interruption in the process yields the wrong cardinal number for the coconuts. Therefore, he does not just plan to begin a sequence of mapping, but plans from the beginning to complete a sequence of steps that will lead him to the end and give him the result he is after.

Where do the various counting acts take place in the brain? To answer this question, it is tempting just to say that both stick-counting and modern counting take place in human consciousness. Find where consciousness is in the brain, and you have found the source of counting. However, when we take this approach we soon run into a dead end. In fact, identifying the specific location within our brains for what we call consciousness has eluded scientists for many years; and after reviewing brain injury cases, many scientists have reached the conclusion that no single part of the brain is responsible for consciousness. In fact, Erich Harth in his book, *Windows on the Mind*, suggests that consciousness is a phenomenon of the entire nervous system.

> The immediacy of our sensations as well as the inability of the isolated cortex to sustain consciousness suggest that the boundaries of consciousness extend well beyond the braincase, probably to the very surface of the body and even beyond.[2]

Since the seat of consciousness does not appear to be localized, finding our counting skills by considering consciousness is of no help. What we need is to take a closer look at the structure of the brain and try to discover where our counting skills might be.

HOW DOES THE BRAIN COUNT?

The brain, of course, is that watery mass of tissue inside our skulls that is responsible for controlling our bodies and for providing a home for our thoughts. Brain tissue is really a soupy gel with a density slightly greater than water, which, if laid out on a table, might rupture under its own weight. The main cells in the brain are approximately twelve billion neurons, or nerve cells (Figure 1). How large is twelve billion? It is more neurons than the number of human beings on the face of the earth (five and a half billion) but less than the number of stars in our Milky Way Galaxy (a hundred billion) and less than our national debt (over four trillion dollars). Each neuron consists of a cell body, a series of branching fibers called dendrites that receive electrical signals from other neurons, and an axon fiber that branches out and carries the cell's electrical signals to other neurons. Considering the connections made between neurons, a single neuron can directly communicate with as many as four thousand or even five thousand other neurons.[3]

Beginning at the spine and moving upward, the brain can be roughly subdivided into three parts: the *brain stem*, the *cerebellum*, and the *neocortex* at the top of the brain (Figure 2). To determine the brain's role in counting, we will concentrate on the neocortex, which is the seat of

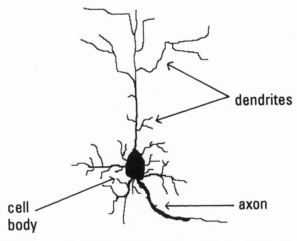

FIGURE 1. A neuron cell of the human brain, which consists of a cell body, dendrites for receiving signals from other neurons, and an axon for transmitting signals to other neurons.

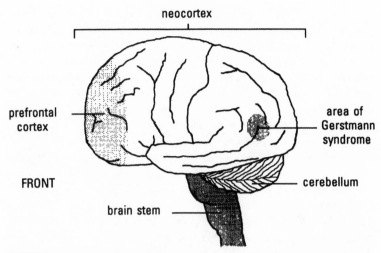

FIGURE 2. The human brain. Both the prefrontal cortex and the Gerstmann syndrome area are intimately connected with counting.

memory, learning, and intellectual skills, and the residence of such important functions as sight, hearing, and language. The outside surface of the neocortex, called the gray matter, consists of a thin layer of approximately ten billion neurons. This gray matter is so densely packed with neurons that one cubic millimeter (which is the approximate size of a pin head) contains from thirty thousand to one hundred thousand neuron bodies. The cortex is divided in half by a large longitudinal fissure running from the forehead to the back of the brain. A bundle of two hundred million axons, called the corpus callosum, connects the two halves of the neocortex together.

It would be nice if we could point to a small bump on the brain and say, "That is where your counting ability resides." Unfortunately, and fortunately, the brain is too complicated an organ for this to be done. Some brain functions seem to be localized in specific areas while others appear to be spread throughout large areas of the neocortex. For example, research has shown that, while many memories are associated with specific areas of the brain, some memories are more general, and can be diminished by an injury without being totally lost. The sharpness of the decrease in memory is consistent with the size of the injury.

Where is the area for mathematics, and where is the specific area or areas for counting? We have no absolute answer to these questions, yet we can make some intelligent guesses. Neurologists, in their attempt to understand how the brain works, have made detailed studies of patients with brain injuries.[4] Even though these studies have sometimes shown contradictory results, we can draw some general conclusions.

The right hemisphere of the brain acts in a more simultaneous manner than the left, and is involved with immediate visual perception, spatial relationships, and movement skills. Children with congenital injury to the right side of the brain sometimes lack the ability to appreciate groups of objects presented to their visual field or to understand the idea of manyness. This ability to appreciate the manyness of things is inherent in humans and not learned. Hence, the right hemisphere appears to be critical for learning to count, since the first step in counting is discovering the manyness of a set.

Even though language is learned, it is associated with specific brain areas. Language is dominated by the left side of the neocortex (though the right side does have some language ability). The left side is also associated with sequential functions such as planning and executing a series of activities. It also dominates symbolic and abstract processes, and therefore is generally credited with mathematical operations, which are both symbolic and sequential in nature. Of special interest is an area on the left side of the brain (Figure 2) which, if damaged, can cause Gerstmann syndrome, a syndrome involving the loss of the patient's ability to identify his or her fingers, causing the loss of fine finger movements. In addition to this finger blindness, the patient has right–left disorientation, and an inability to do simple arithmetic calculations.

At first the association between recognizing our fingers and performing arithmetic calculations may seem strange. However, as babies, we learned to count by having someone point to each of our fingers while saying a number word. Hence, we have a strong association between our fingers, counting, the sound of number words, and the manipulative skill of pointing from one finger to the next.

Some neurologists[5] have suggested that counting and computational skills reside in an area at the very front of our brains called the prefrontal cortex. This part of the brain helps control emotions and gives us a sense of responsibility, a sense of the past and future and the ability to plan future activities. Some prefrontal injury patients have lost their calculating abilities. Counting involves a planned, manipulative activity, and therefore any damage to this area could interfere with such activity.

What is the final conclusion about where counting takes place in the brain? There is no sure answer. What is certain is that different areas are involved in the collection of skills we know as number awareness, counting, and calculating. The right brain recognizes multiple objects in our immediate visual field to give us a sense of manyness. The prefrontal cortex enables us to plan counting operations. In the back of the left hemisphere, we associate our fingers with counting. While this is happening, we incorporate the language parts of our left brain to bring forth the number words. Using these various brain functions, we point to each finger and speak each number word in its correct order.

HOW DO WE LEARN TO COUNT?

We began this chapter by considering what numbers are and how we use them. We have wandered the terrain of the human brain but have not come up with any simple answer to how we count. Still, we have certainly improved our understanding of the complex nature of our counting and calculating talents. In conclusion we will consider the genesis of how we acquire the skill of counting.

When we are born, all of the synapses or connections in our brains have not been made. The process of neurons connecting with each other continues into our sixth year. Hence, growing up involves millions or billions of nerve cells shooting out their axons to connect with other nerve cells. Initially, our brain functions are not strongly divided between the left and right side. It is only after an appropriate amount of growth that certain functions settle in one side or the other. While this growth is taking place, our parents are trying to teach us to count. Obviously, when we are only a few weeks old we cannot learn this because our brains are not big enough and too few neurons have made their final connections. But around the age of two or two and a half, we begin to put the different parts together that will lead to counting and a sense of number.

We begin because of two conditions. First, our motor and verbal skills have developed enough for us to sit and listen to our parents. Second, one of them sits on the floor with us and, taking one of our hands, says: "Look, these are your fingers." The parent then proceeds to touch each finger and say the appropriate number word. "One, two, three. Now, you do it." This is a learned behavior because, if no one does this for us, we do not learn to count. As noted earlier, certain primitive tribes do not count beyond the number two since their culture does not require it.

When we learn to count—between the ages of two and four—it is a struggle and not something we pick up all at once. It is a struggle mainly because our brains are still growing and because worried parents may betray their fear that we may not measure up intellectually. Hence, they push us to the limits of our abilities (as even many loving parents do). At first, counting is a very deliberate act for us as we labor to remember the correct number word and say it as we point to the correct finger. Such deliberate activity may well be controlled by the prefrontal cortex. After sufficient practice, we memorize the number words as a distinct subdivision of our vocabulary, and create the associative long-term memories for each number. These associative memories include sound, visual, and movement memories for each number. In addition, each number has the associative memory of the next number of the sequence (e.g., when I say "three," an automatic association is the next number "four"). At first, all these memories are strong, but in time the sound memories become dominant. At this point we have the correct spoken number words, but written number words and arabic numerals generally come after we begin school. Movement memories associated with numbers come from our finger activity—pointing. Even as adults this association remains strong, for we frequently and subconsciously touch our thumbs to our other fingers as we mentally count. When counting is finally automatic, these long-term associative memories are located in different sites on the neocortex.

Even though we learn how to count, we can see that the brain functions that allow us to learn counting are part of our biological inheritance. In addition, the very first act of counting, asking how many objects are present in a visual field, appears to be inherited and available to all humans. Therefore, the basic question of "how many" that counting (and numbers) are meant to answer is a question buried in our very natures as human beings.

CHAPTER 2

Early Counting

We will now investigate how long people have been counting. If counting is a recently acquired skill for human beings, then we might conclude that it is not, after all, intimately connected to our basic nature. If, on the other hand, we find that it is ancient, dating even as far back as our prehuman ancestors, we could conclude that counting is part of what it is to be human, just as language and tool-making are fundamental to our species.

Counting is much older than recorded history, and we cannot say that it was on July 2, 1103 B.C., along the Nile river that Rajamman the Elder invented counting and taught it to his village; counting is much older than that. With its origins cloaked in prehistory, we must reconstruct our evolutionary past and then guess when the skill of counting might have originated. We will do this for both forms of counting: stick-counting, which relies on a manipulative technique to find the cardinal number for a set; and modern counting, which uses number words.

OUR DISTANT ANCESTORS

Our knowledge of the evolutionary history of human beings is still developing, and many anthropologists have conflicting ideas of just how this history progressed. Apes, including the gorillas, chimpanzees, orangutans, and gibbons, appear to be our closest relatives still surviving on the planet. Of these four, it is the chimpanzee which is generally considered to be our closest relative. Allan Wilson and Marie-Claire King from the University of California at Berkeley have compared the proteins of humans and chimpanzees and found the difference to be only 1%, leaving 99% of the proteins the same. It is also the chimpanzee that shows a high degree of general intelligence by recognizing itself as a separate and distinct being. In tests to find out how quickly apes can recognize their own images in a mirror, the chimpanzee is the quickest, followed by the orangutan, and in third place is the poor gorilla, who fails the test entirely.

Human beings comprise the only known species that can truly count, that is, assign the correct cardinal number to a set using either stick-counting or number word counting. Therefore, counting must have originated later than the time when the evolutionary line branched into human beings and apes. On the opposite side, counting is older than written language, which originated in the Middle East about five thousand years ago. Hence, the window when counting began is bracketed between the time the human line broke from the apes and 3000 B.C. This time gap, as we will see, is very large.

Allan Wilson and Vincent Sarich (also at Berkeley) conducted studies of the similarities and differences between ape and human proteins in an effort to determine how long ago that evolutionary branch occurred.[1] Their conclusion, and the one many anthropologists accept now, is that the split between apes and man occurred from five to six million years ago. Since that time, our brains have tripled in size compared to the chimp's brain, and the number of neurons has doubled. Hence, the major difference is the size of our brain, especially the size of our neocortex, and, as we said earlier, it is the neocortex that is the key to understanding counting.

Who is next in our evolutionary line after our common ancestor with the apes? The discoveries of fossil fragments of animals that look something like apes, yet have similarities to human beings, are so numerous that we simply cannot review all their origins. Fortunately, anthropologists have classified them into several groups, each group having distinct characteristics that illustrate the steady march from animal intelligence to modern human intelligence. All of these apelike, humanlike animals are called hominids.

The oldest hominids are the African *Australopithecines*, whose fossils have been dated from approximately four million years to one and a half million years ago. Therefore, this fellow lasted a long time—two and a half million years. It would be to our credit to match that record.

What were the *Australopithecines* like? The major characteristic that separates them from modern apes and monkeys is their ability to walk upright in an erect posture, which was the first milestone toward becoming human. They were small—well under five feet—and under ninety pounds in weight, probably shared their food within a family and local group, had a nuclear family structure, and may have used a division of labor based on gender. However, with all of this, they still had a brain approximately the same size as an ape's—an average of 450-cm^3 compared to the chimpanzee's 400-cm^3 brain and the gorilla's 500-cm^3 brain. No convincing evidence exists that *Australopithecines* manufactured tools, ate meat, used fire, made clothing, buried their dead, or created symbolic art—all characteristics of later hominids.[2] The most important feature missing from *Australopithecines* is a large brain. This strongly suggests that they did not have the capacity to conceptualize the question of "how many" and then create a plan to find a solution. Hence, we may conclude that they did not count.

Approximately two million years ago, and lasting for a half-million years, a new hominid appeared in East Africa, *Homo habilis*, or "handy man," (*homo* is Latin for "man"). They are called this for a very good reason: They manufactured stone tools. This represented a major advance over the older *Australopithecines*. In addition, their brains were larger than the *Australopithecines*, averaging 750 cm^3 as opposed to 450 cm^3. This is a substantial increase, yet in most other ways, the *Homo habilis* were much like the earlier *Australopithecines*. They were under five feet in height and averaged about ninety pounds in weight. Examination of their teeth shows that they probably ate fruit like the *Australopithecines*, and were not omnivorous as were later hominids. We still do not see the use of fire, clothes, or art. However, a larger brain and the manufacture of tools represented giant steps forward.

Is there any evidence to suggest that *Homo habilis* used some form of counting? Not really. *Homo habilis*, like their predecessors, stayed in East or South Africa where the constant climate permitted food gathering year round.[3] They ate fruits and had no need to accumulate stores of food over time. There is no evidence that they stayed in permanent camps, which would be associated with the accumulation of possessions. Without possessions, and secure with a constant food source, they probably had no need for counting. Even though they had a larger brain than *Australopithecines*, there is no evidence they had rapid phonetic speech.

Around one and a half million years ago evolution took an interesting turn with the appearance of *Homo erectus*, or "upright man." He survived until about 300,000 years ago. *Homo erectus* represented a marked advance over the previous hominids. The following attributes describe him:

- Had a much larger brain than *Homo habilis*
- Had more sophisticated tools than *Homo habilis*
- Used fire
- Migrated out of Africa into Europe and Asia
- Had seasonal, semipermanent base camps
- Constructed shelters
- Was both plant- and meat-eater (omnivorous)

From the above list you can see the dramatic advance in *Homo erectus* over both *Australopithecines* and *Homo habilis*. The increase in the size of their brains was substantial. Because *Homo erectus* changed markedly over the 1.2 million years of its existence, this culture has been divided into early *Homo erectus* and late *Homo erectus* periods. Early *Homo erectus* had an average brain size of 900 cm^3 compared to *Homo habilis* with 750 cm^3 (Figure 3). The brain of later *Homo erectus* increased to 1100 cm^3, which is approaching the average for modern humans, 1400 cm^3.

The tool manufacturing of *Homo erectus* showed improvement over that of *Homo habilis* and was so sophisticated that modern anthropologists, attempting to imitate their abilities, required months of practice to become skilled at making *Homo erectus* tools. Evidence exists that *Homo erectus* constructed simple shelters, a skill lacking in *Homo habilis*. Yet, the most significant advance in technology that *Homo erectus* achieved was the domestication of fire, which probably occurred over a period of

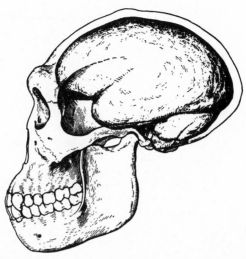

FIGURE 3. Skull and brain case of *Pithecanthropus erectus*, first *Homo erectus* discovered in 1891 by Eugene Dubois on the island of Java. The skull is approximately 450,000 years old, and the brain case has a volume of 900 cm³. (Drawing from Stock Montage, Chicago, IL.)

half a million years, from one and a half million years ago to one million years ago. While *Homo erectus* was mastering the use of fire, he also moved out of Africa into seasonal climates in Europe and Asia.

Even his appearance resembled modern human beings more than his predecessors' had, for *Homo erectus* was taller than the *Homo habilis*, averaging over five feet, with a much flatter face. His head was bigger and more rounded, yet it still had pronounced ridges over its brows and a sloped forehead. If you saw one dressed in a business suit, waiting for a train, you might take a second hard look, but you would eventually pass him by. In contrast, if you saw *Australopithecus* or *Homo habilis* standing on the platform, you'd probably call Animal Control and say there was an ape loose.

The migration of *Homo erectus* from Africa to Europe and Asia was nade possible by his use of fire, construction of shelters, and omnivorous liet. In severe climates, fires and shelters helped him survive the cold winters. In addition, when fruits and vegetables disappeared in the fall, he could hunt meat and keep it refrigerated out of doors. Without access to fire, shelter, and meat, it is doubtful any hominid could have survived beyond the secure warmth of Africa.

Did *Homo erectus* count? There is no evidence that *Homo erectus* had rapid phonetic language, therefore, he did not master modern number word counting. But did he have stick-counting? This is a more difficult question. It is certainly not necessary to know language to have a number sense, for both animals and human beings have it. David E. Smith comments in *History of Mathematics* that

> A well-known instance of a deaf-and-dumb boy who acquired a knowledge of numbers from observing his fingers, even before he was taught to count, shows us that the idea of number did not have to await the development of spoken language, and so a savage may appreciate three without having a name for numbers beyond two.[4]

There is no direct physical evidence that *Homo erectus* did use stick-counting. That is, archaeologists have not found numbered markings on their tools or numbered patterns of stones on the ground at their camp-sites. What indirect evidence suggests stick-counting for *Homo erectus*? First, they had large brains that, especially in later *Homo erectus*, approached the size of the modern human brain. Second, they achieved sophisticated tool manufacturing. To make a tool, one must plan a sequence of actions that will result in a desired end product. This, of course, is just the kind of brain function needed for mapping sticks, nuts, or pebbles onto a set of objects.

Homo erectus moved out of the warm climate of Africa and into the seasonal climate of Europe and Asia. With a cold winter to survive, *Homo erectus* certainly stockpiled food resources. This may have been the first environmental pressure to encourage stick-counting. Other problems may also have encouraged the development of counting. How many days of travel to the next community? How many days until the full moon (when they could hunt at night)? How many hunters are left after being chased by bears?

Without direct proof, we are going to speculate that *Homo erectus* probably developed stick-counting. This would have been in conjunction with their larger brains, better tools, use of fire, and migration to cold climates. It's unlikely that stick-counting evolved throughout all tribes of *Homo erectus* at once; it probably evolved erratically. Perhaps it evolved in several places only to die out and then begin again. We may one day solve this riddle if some lucky archaeologist stumbles across a *Homo erectus* site that contains hard evidence. Until that time we will have to ponder the question.

About three hundred thousand years ago, *Homo erectus* began disappearing. Before this time, maybe as early as five hundred thousand years ago, new hominids appeared in Europe, Africa, and Asia.[5] They were *Homo sapiens* or "wise man." *Homo sapiens* represent a major change over *Homo erectus*: the brain size increased to approximate the modern size of 1400 cm[3]. However, *Homo sapiens* were not modern human beings. They were a transition from *Homo erectus* to us, but they did not have the full rounded cranium, high forehead, and flat face characteristic of modern people.

If we are correct that *Homo erectus* used stick-counting, then certainly *Homo sapiens* used it too. In either case, the larger and more complex brain of the *Homo sapiens* would have been more conducive to the development of such mental skills as stick-counting. Since *Homo sapiens* remains are found in Europe and Asia, as well as Africa, they, too, had to survive in colder climates. Did they use modern language? The opinions of the experts vary on this, but the consensus is that they did not use our rapid, phonetic speech. However, there is evidence that they did use some sort of protospeech based on a limited range of vocal sounds.[6]

Now we finally reach the last stage of our evolution: the appearance of modern human beings called *Homo sapiens sapiens*. The final changes included our large brain and a fully articulated larynx and throat for rapid speech. In addition to making complex tools, we bury our dead and create art.

When did *Homo sapiens sapiens* first appear? There is some controversy over this point. Archaeologists have uncovered bones at sites going back at least thirty thousand or thirty-five thousand years. However, recent evidence has accumulated that suggests modern human beings are much older. Using two new controversial techniques for dating hominid sites, archaeologists have determined that some sites are as old as a hundred thousand years. If these ages hold up, then *Homo sapiens sapiens* have been around for a hundred thousand years and not just thirty-five thousand years.

Our best guess at this time is that *Homo sapiens* evolved from five hundred thousand to a hundred thousand years ago into modern people. Scientists speculate that modern language evolved during the past hundred thousand years. It probably evolved from earlier song rituals performed around the campfires, and was a means of retelling the tribe's history and culture. In all likelihood, number word counting developed sometime after we began to use modern speech—about a hundred thousand years ago.

FINGER- AND BODY-COUNTING

Before moving on to language counting, we must consider a kind of counting that originally grew out of stick-counting, or possibly in association with stick-counting. Instead of using sticks, pebbles, or shells, the counter used his or her own fingers in much the same fashion as does a modern child. One advantage, of course, is that fingers are convenient. Finger-counting also has two substantial disadvantages: fingers are limited in number (even if you add the ten toes) and do not work well as a record. If you hold up eight fingers formed from a finger count, it's inconvenient to walk around all day holding these same fingers up as a record of what you have counted. It is much better to count with pebbles and then put them aside, to be referred to later. (We are still assuming here that the individual user has no abstract idea of number, so it is not possible for that individual to "remember" the number "eight" for future reference.)

Finger-counting is universal, for in every part of the world people have used some form of finger-counting in the past. Even after the modern number sequence was named and in use, societies still used finger-counting. The Romans used it to count up to ten thousand. Even today, evolved forms of finger-counting and finger-calculating are still used, primarily by ordinary men and women rather than the scientifically elite. Since it is nonverbal, it may be as old as stick-counting. But we are interested in finger-counting beyond its association with stick-counting, for it represents a significant advance in number concept. Earlier we defined two kinds of numbers: cardinal and ordinal numbers. A cardinal number represents the number of elements in a set, while an ordinal number shows its order within the set. In simple stick-counting we get only cardinal numbers. The final pile of stones or bundle of sticks represents the cardinal number of the set counted, but we get no sense of order. We find no ordinal numbers because the sticks and pebbles are treated as identical to each other and are interchangeable.

However, when we count with our fingers we potentially add order to our count. The fingers are distinct, with distinct names, and, when used in counting, they are counted in a definite order. For example, in many parts of the world people begin with the little finger of the left hand as the first counting finger, then proceed to the ring finger, and then the middle finger. This suggests ordering since the ring finger comes after the little finger but before the middle finger, just as the number two is after one but before three. Hence, finger-counting offers the opportunity to introduce ordinal numbers into our number concept. Since we can finger-count in a definite order, our numbers inherit a definite order, and the number sequence is born.

From finger-counting grew body-counting. Here, the one doing the counting points first to fingers and then points to other parts of the body as the number count increases. Remnants of old body-counting are still found in parts of Australia and New Guinea. A tribe on the island of New Guinea has the elaborate body-counting scheme presented in Table 1.[7]

The utility of body-counting is apparent at once when we consider the special requirements of coordinating a group of hunters or warriors. The ability to both signal commands and pass information over distances without the use of sound is of great advantage when sneaking up on game or one's enemies.

Finger- and body-counting were once worldwide and are certainly very old. Their provision for ordinal as well as cardinal numbers is a substantial move toward the abstraction of true numbers. If we ask a farmer how many apples he has in his basket, he may answer by holding up a bundle of sticks. By just looking at the bundle, we cannot tell the cardinal number of apples; the best we can do is get a feel for their

TABLE 1.   Example of a Papua Island
Native Body-Counting Scheme

1 = right little finger	12 = nose
2 = right ring finger	13 = mouth
3 = right middle finger	14 = left ear
4 = right index finger	15 = left shoulder
5 = right thumb	16 = left elbow
6 = right wrist	17 = left wrist
7 = right elbow	18 = left thumb
8 = right shoulder	19 = left index finger
9 = right ear	20 = left middle finger
10 = right eye	21 = left ring finger
11 = left eye	22 = left little finger

Data from Karl Menninger, *Number Words and Number Symbols* (New York: Dover Publications, 1969), p. 35.

magnitude. However, if he points to a specific part of his body, as in body-counting, then that single gesture gives us the exact cardinal number. We may not have a name for that number, but we immediately know what it is. By using body-counting he has found and communicated to us the ordinal value for the number. Hence, we know the relative position of the number in relation to other numbers, that is, in relation to other body parts that represent numbers.

COUNTING WITH WORDS

Counting with number words is an added degree of abstraction for human beings. While using stick-counting, and even finger-counting, we do not have true abstract numbers, because our process is mechanical and does not require abstract thinking. On the other hand, abstract objects are objects of thought, and the natural numbers are perfect examples of such abstract entities. But if all we are doing is manipulating sticks to compare them to other physical objects in a set, then we do not have to think abstractly. But as soon as we make a sound to represent a number, then the abstraction has begun. Just as we abstract the name "dog" to apply to any of a number of hairy, four-legged beasts who always want to be fed and scratched, we can abstract "two" to mean any set that has a pair of elements. We then use the sound "two" to refer to such a set.

When did human beings finally decide to assign words to numbers instead of using sticks, shells, or stones? Probably not before a hundred thousand years ago since this is when rapid phonetic language first appeared. Yet, number words did not suddenly pop into the air as soon as people began using language. The development of number words was slow, advancing in gradual stages to the present decimal system. In fact, we might suppose that the evolution of number words was delayed because *Homo sapiens sapiens* already possessed perfectly good methods of determining the cardinal numbers of sets through stick- or finger-counting. Why would they need words?

To get some idea of how the earliest number words evolved, let's look at the numbering systems of some existing primitive tribes that have been studied by anthropologists. As we have noted, certain tribes, like the Australian aborigines, when first encountered by modern Europeans, could not verbally count beyond two.[8] They had a word for "one" and a word for "two" but everything else was simply "many." The Bergdama tribe of South Africa also had only the two number words, "one," and "two," with everything else assigned to "many." Whether these tribes could stick-count has not been recorded. We must be careful not to assume that a tribe's total understanding and use of numbers are confined to the number of number words in their language. As Graham Flegg states in *Numbers: Their History and Meaning*:

> Here, it is important to sound a certain note of warning. We must not jump to the conclusion, as some anthropologists have done, that people have no appreciation of numbers beyond the limit of their number-words. Words are very often found to be accompanied by gestures. Both must be properly investigated before general conclusions are drawn.[9]

Many other primitive tribes, from Africa to South America to New Guinea, had only two number words, one and two, but combined them to count larger groups in what is known as 2-counting. In 2-counting we simply repeat the number words for one and two the appropriate number of times. An example of 2-counting is the Gumulgal tribe of Australia.[10]

Their word for one was "urapon" and their word for two was "ukasar." They counted in the following fashion:

1 = urapon ("one")
2 = ukasar ("two")
3 = ukasar-urapon ("two-one")
4 = ukasar-ukasar ("two-two")
5 = ukasar-ukasar-urapon ("two-two-one")
6 = ukasar-ukasar-ukasar ("two-two-two")

The system of 2-counting used by these tribes could go on to higher numbers, but they generally did not, because the string of words needed for higher numbers soon becomes too difficult to remember. The user of a 2-counting system frequently accompanies the verbal count with a finger count. Hence, the 2-counting system is very restricted and of little utility, since it can be easily replaced by simply counting on the fingers of both hands. In fact, 2-counting with words is far less efficient and useful than stick-counting. With stick-counting the user not only can count to higher numbers, but he or she has a semipermanent record of the count in the bundle of sticks or pile of shells. In verbal counting, the count seems to evaporate into the air soon after we say the last number word.

Because of the diverse sites where 2-counting has been found, from Africa to South America to Australia, a debate has ensued whether 2-counting evolved separately on each continent, or in one place and then spread around the globe. In either case, it is certainly old, possibly as old as language itself. A later development was a sophistication of 2-counting to include multiplying certain numbers to get larger numbers. For example, three different words might be assigned to "one," "two," and "three," and then "six" would be "two-threes" which means two multiplied by three instead of two plus three. This is called neo-2-counting, which has been found in greater abundance around the world than pure 2-counting, but frequently in locations adjacent to pure 2-counting.

The next phase in the evolution of number words was to 5-counting, which must have been suggested by the number of fingers on one hand. In fact, in some primitive languages that utilize 5-counting, the number words beyond four actually describe the hand gestures used to finger-count the corresponding number. Versions of 5-counting have been found on most continents. In one South American system we get:

five = "whole-hand"
six = "one-on-the-other-hand"
ten = "two-whole-hands"
and eleven = "one-on-the-foot"

By the time we get to twenty we arrive at "one-man," indicating all ten fingers and ten toes.[11]

The 5-count system evolved into two different kinds of number-word counting: 5–10 counting and 5–20 counting. The 5–20 system may actually be older. Table 2 gives a generalized scheme for each type. Only the numbers in increments of five are shown; the intervening numbers would have similar add-on number words.

In actual practice, the two systems mixed in different combinations. The words used were frequently names or combinations of names of individual fingers, parts of the body, or even various counting gestures. We see vestiges of this in our modern counting system where individual numbers are "digits," which comes from the Latin "digiti" meaning fingers.

NUMBERS AS ATTRIBUTES

When numbers were first assigned word names, they did not have the abstract meaning they have today. In the beginning they were used as adjectives to describe the physical objects being counted. We still see a

TABLE 2.   Generalized 5–10 and 5–20
Counting Schemes

	5–10 Counting	5–20 Counting
10 =	"ten"	"two fives"
15 =	"ten and five"	"three fives"
20 =	"two tens"	"twenty"
25 =	"two tens and five"	"twenty and five"
30 =	"three tens"	"twenty and two fives"
35 =	"three tens and five"	"twenty and three fives"
40 =	"four tens"	"two twenties"

remnant of this in modern language. We speak of a yoke of oxen, and know at once that two oxen are meant. We would never say "a yoke of gloves." Yoke, as a number adjective, does not apply to gloves. We have a pair of gloves, but a "pair of people" sounds a little strange to the ear. The correct number word for two people is generally "couple." Indigenous people frequently use different number words for different objects. In the Fiji Islands, natives use the word "bola" for 10 boats, but "koro" for 10 coconuts.[12] At one time a Native American tribe from British Columbia used seven different classes of number names for counting different kinds of objects.[13] The use of different number-word classes in different languages demonstrates the close connection early human beings saw between numbers and the objects counted. In *Number Words and Number Symbols*, Karl Menninger explains:

> The number classes once more reveal clearly how closely the number was involved with the object in early man's conception, how strongly things dominated numbers. They also illustrate the mental obstacles that primitive man had to overcome not only in liberating the number sequence from these preliminary stages but also in building upon the earliest beginnings.[14]

Slowly, the number words became generalized and could be used to count an expanding collection of objects. This made possible the contemplation of numbers as purely abstract objects not associated with any physical entity. The 5–10 and 5–20 systems gave way to the 10-system we use today. In rare cases, other numbering systems survived for a time, only to die out. There is evidence for a 12-based system, and the Sumerians and Babylonians used a 60-based system. Yet, the majority of cultures across the globe now use the 10-system of counting.

NUMBERS BEFORE FARMING

For most of our history, people lived in caves or huts, hunted buffalo with spears, and gathered fruit, nuts, and roots. This lifestyle has been aptly named "hunting–gathering," for the women gathered plant foods while the men hunted animals. For the last hundred thousand years they probably used stick- and finger-counting, since all the skills associated with modern human beings were by then fully developed, and survival pressures would certainly have encouraged such counting. Just when did the 2- and 5- number-word systems evolve from this earlier counting?

We can assume that hunting–gathering was the mode of life for humans from the time of *Homo erectus*, the first group to adopt an omnivorous diet, until farming was invented eleven thousand years ago. This would cover a time span of one and a half million years. Hunting–gathering must have been a successful means of survival to have lasted for such a long time.

When we imagine how people lived as hunter–gatherers, we are tempted to envision a small band of dirty people dressed in skins and near starvation as they trudged through the wilderness searching for their next meal. However, studies of modern hunter–gatherers reveal that many bands have more free time than their counterparts in urban society, and that they do not experience the constant threat of starvation.[15] In fact, if our hunter–gatherer ancestors had always lived on the edge of extinction, how could their lifestyle have prospered for one and a half million years and spread over the entire globe?

A popular hypothesis exists that once people gathered in cities and became "civilized," spare time was abundant. With this spare time, the rulers, priests, or civil servants invented writing, mathematics, and science. If this were so, why did not the hunter–gatherers, some of whom must have spent much time sitting around discussing the world, invent them? They did not because the hypothesis is false. Spare time does not inspire people to invent things, problems do. Writing, mathematics, and science were invented because people encountered problems whose solutions required them.

True farming, that is, the deliberate and selective planting of seeds for growing food, began approximately eleven thousand years ago in the Fertile Crescent. What was happening before this time? The oldest direct evidence we have for counting is a thirty-thousand-year-old wolf bone found in Czechoslovakia in 1937 by Dr. Karl Absolon.[16] A drawing of the bone is shown in Figure 4. It has fifty-five notches carved in it, grouped in groups of five—clearly a deliberate record of counting. This wolf bone suggests the use of a 5-counting system. Hence, prehistoric hunter–

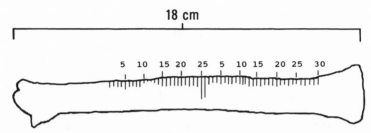

FIGURE 4. Drawing of a thirty-thousand-year-old wolf bone found in Czechoslovakia in 1937 by Dr. Karl Absolon. The fifty-five grouped notches demonstrate that counting was a talent possessed by its maker. [Drawing adapted from photograph, Lucas Bunt, Phillip Jones, Jack Bedient, *The Historical Roots of Elementary Mathematics* (New York: Dover Publications, p. 2.)]

gatherers may have developed the earlier 2-counting and neo-2-counting systems in addition to a 5-counting system.

Pure 2-counting uses the principle of addition. The number four is "two-two" which means, of course, two plus two. Six is "two-two-two" which is just two plus two plus two. Therefore, it would seem that the hunter–gatherers who lived more than thirty thousand years ago understood the principle of adding small natural numbers since 2-counting preceded 5-counting. We can also guess that they had at least a rudimentary understanding of multiplication because the neo-2-counting, which came after pure 2-counting and before 5-counting, involved the multiplication of small natural numbers. We see this in the number for six, which was generally "two-three," meaning two multiplied by three and not two plus three. Since the Czechoslovakian tribe that created the wolf bone was probably 5-counting, we can assume some human beings at that time had progressed beyond the neo-2-counting stage. This supports the belief that Upper Paleolithic Age hunter–gatherers (who lived seventy thousand to twenty thousand years ago) understood simple addition and multiplication.

Subtraction is also suggested, since some early neo-2-counting systems probably used subtraction to form the numbers seven and nine as was done by some tribes into the twentieth century. Certain tribes form seven as $2 \times 4 - 1$ and nine as $2 \times 5 - 1$.[17] The evidence for division is even more indirect. Yet, considering how important the distribution of food must have been to all early human beings all the way back to the *Homo erectus*, the principle of division must have been understood for at least the simpler fractions. The concept of dividing something in half, in quarters, and in thirds must have been known. It would have been incongruous for people who could fashion complex spear points, organize coordinated hunts, and survive in cold winter weather to have had no idea of how to divide something in half. Yet, knowledge of the process of simple division would not imply that hunter–gatherers had developed the idea of fractions as numbers. In fact, such evidence is lacking. The physical dividing of something into two halves may have been regarded as changing that something into two new items, not the creation of two "one-halves." Having no evidence to the contrary, we should assume for now that the ancient hunter–gatherers had not developed the concept of any number beyond the natural numbers.

Based on the above evidence, what would our best guess be for the evolution of numbers prior to farming, eleven thousand years ago? In all likelihood, most human beings probably recognized the first natural numbers. At least some peoples not only could count very proficiently using a 5-counting system but also could probably manipulate small natural numbers in the operations of arithmetic: addition, subtraction, multiplication, and maybe even division. We must be cautious and not infer too much mathematical ability for the hunter–gatherers. Many hunter–gatherers who have survived to the modern era are generally limited to the more rudimentary neo-2-counting system. However, it is still possible that new archaeological finds, or reinterpretation of old finds, will shed additional light on the mathematical understanding of our prefarming ancestors.

PUTTING IT ALL TOGETHER

We have uncovered a number of steps which, as best as we can surmise, led from primitive stick-counting to modern number words and number systems. Although direct evidence is often lacking, we can speculate on the progression of steps in the understanding of the natural numbers. It took a million years for our ancestors to abstract the idea of numbers. Hence, the abstraction of numbers was excruciatingly slow. Yet, this scenario also supports the view that counting was not a modern development, invented by Egyptian priests who had too much time on their hands, but was a skill acquired over many millennia, a skill that may, in some form, be as old as our use of fire.

We are tempted to conclude that we of the twentieth century are wise and clever, while our poor ancestors were dullards. Why did they take so long to develop numbers and counting? Why did they not just smarten up and conceive of abstract numbers right away? What this conclusion ignores is that our ancestors, especially over the last hundred thousand years, were as smart and as clever as we are. A great range of intelligence certainly must have existed among people then, just as it does today. Some were most definitely on the slower side, yet others were probably brilliant. The level of craftsmanship and ingenuity in some of the ancient buildings and monuments, such as Stonehenge and the pyramids, shows that intelligent people have lived in every age of *Homo sapiens sapiens*. If, in ages long past, brilliant minds labored over the idea of numbers and counting, then we are not brighter than our ancestors because we learned numbers as children. We simply enjoy the benefits of inheriting what it required many generations to develop.

The march toward a concept of numbers began very slowly, the first steps spreading over hundreds of thousands of years. As we progressed in our development of a number system, the major steps came ever faster. This trend continues with the next phase of our journey into the history of written numbers at the dawn of civilization.

CHAPTER 3

Counting in Other Species
How Smart Are They?

Nothing causes an argument quicker than a discussion about the intelligence of animals; dog lovers attack cat lovers, horse enthusiasts admire their icon, while pig fans sit back and smile at all the confusion. The question of whether or not animals can count is just as divisive. Some scientists and most pet owners claim that, not only can their favorite pets count, they can add, subtract and, on special occasions, read their masters' minds. "How did Sparky know I was going for his leash?" says the devoted dog owner. "I could have been getting up to go to the kitchen. But he knew. He jumped up and ran to the front door. The only explanation is that he read my mind."

Our question is more specific than the general intelligence of animals; we concede that some appear to be rather smart, while others seem plain stupid. What we want to know is whether any animals can count, and hence, have a concept of number. Our earlier definition of modern counting involved mapping number words onto the elements of a set. Since human beings are the only living species known to use phonetic, rapid-speech language, we realize that this definition excludes all animals from counting in this manner (with the possible exception of dolphins and whales—animals that do seem to have rapid sound communication). But, maybe this definition is too restrictive. What we really want to know is whether any animals can

1. recognize the manyness of a set,
2. have a desire (or need) to know a set's multiplicity,
3. by mapping, compute the cardinal number of a set.

The above three requirements are consistent with our definition of stick-counting, which eliminates the need for language to denote a set's cardinal number. Even with this looser definition of counting, experts disagree on what animals can do. Graham Flegg, a founder–member of the Open University as Reader of Mathematics and past president of the British Society for the History of Mathematics, comments regarding stories of animal counting:

> These and other, similar stories have led some people to infer wrongly that creatures other than man can actually count. . . . To the question "can animals count?" we are bound to reply a firm negative.[1]

Yet, on the other side of the question we have the opinion of Professor H. Kalmus, University College of London:

> There is now little doubt that some animals such as squirrels or parrots can be trained to count . . . . counting faculties have been reported for seals, rats and for pollinating insects. Some of these animals and others can distinguish numbers in otherwise similar visual patterns, while others can be trained to recognize and even to produce sequences of acoustic signals. A few can even be trained to tap out the numbers of elements (dots) in a visual pattern. . . . The lack of the spoken numeral and the written symbol makes many people reluctant to accept animals as mathematicians.[2]

How can two experts in the field of animal behavior reach such conflicting conclusions? What is the answer—can animals count or not? To help us with this question, let's review some of the more intriguing cases.

SPECIAL CASES

Almost everyone has heard the story of Clever Hans, a horse who could, at the command of his trainer, not only count, but add and subtract. Hans deftly pawed out the correct answers to mathematical questions put to him. Unfortunately, investigators discovered that the horse was actually reacting to unconscious signals from his trainer to stop his pawing at the appropriate place. In reality he could neither count nor do arithmetic. Poor Hans. This case put researchers on the alert for other instances where trainers or scientists inadvertently gave signals that observant animals could detect in order to "solve" the problems put to them. In fact, research mistakes based on such miscues have been given the name of Clever Hans errors. With Clever Hans in mind, researchers are ever on the lookout for poor research or exaggerated claims.

The next case is less well known but more intriguing. As reported by Sir John Lubbock, a nineteenth-century astronomer and mathematician, an estate owner was bothered by a pesky crow that kept nesting in his watchtower.[3] If the man entered the tower to dispatch the crow, the bird simply flew outside and remained there until the man left. To deceive the crow, the owner sent two men into the watchtower and had one leave. But the bird was too smart and remained outside. The next day the man repeated the operation, but with three men entering and two leaving. Still the bird would not return to the tower to meet his fate. Finally, when five entered and four left, the crow was tricked and returned to the watchtower. Here the story ends and we must rely upon our own imaginations to decide if the crow was sent to crow heaven. What are we to conclude from this story? Presumably, a crow could count to four and not to five. Yet, if the implication is correct, then possibly crows can count.

Formal bird studies confirm the above evidence that birds appear to count. Otto Koehler, past professor of zoology at the University of Freiburg, conducted experiments in which birds were trained to recognize various numbers of dots.[4] A raven was trained to select the appropriate pattern from two to seven dots. Koehler concluded that birds can recognize number patterns from two to seven with what he called "unnamed numbers," even though he did not call this counting, since he maintained true counting involves language.

> Our birds did not count, for they lack words. They could not name the numbers that they are able to perceive and to act upon, but in actual fact they learn to "think unnamed numbers."[5]

Many have heard of the wonderful dances performed by bees to tell their companions where pollen can be found. An insect story more closely related to counting is told by Levi Leonard Conant, past mathematics professor and president of the Worcester Polytechnic Institute.[6] Wasps, before sealing up their eggs in cells, place other dead insects inside as food for the young wasps when they hatch. Interestingly, each species of wasp puts in a constant number of food victims, some five, others ten, and one species stuffs in twenty-four. Experiments have demonstrated that they do not add victims until the cell is full and use this as a signal to stop. How do they know when they have reached the right number? How do they know that exactly ten or twenty-four is it?

In one species, the female wasps grow much larger than the males, so the mother wasp places ten insect bodies with the female egg, but only five with the male egg. How does she know to do this? Is it all instinct? We are faced with the following: if lowly wasps, with their primitive nervous systems, are counting, then most animals probably have the ability to count. On the other hand, if this amazing feat is due to instinct alone, then instinct can mimic the behavior of counting, and we must be doubly on our guard against misinterpreting the actions of animals.

We have discussed birds and insects, but what about mammals? Guy Woodruff of the Primate Facility, and David Premack, Department of Psychology, both of the University of Pennsylvania, have conducted counting research with chimpanzees. They discovered that chimpanzees not only can identify multiple objects from one to four but also can correctly identify proportions of one-fourth, one-half, three-fourths, and one. Yet, Woodruff and Premack are not ready to claim the chimps were counting, ". . . and although the ape may yet be taught to count, this has not been achieved so far."[7]

In research conducted with dolphins, Louis M. Herman determined that they could remember the correct order of a string of up to eight abstract symbols.[8] This is beyond the normal range for human beings who generally fizzle out after six or seven symbols (our phone numbers are separated by a dash between the third and fourth digits to facilitate the memorization of only seven digits). Even though memorizing a sequence of eight abstract symbols is impressive, it is not counting.

Where does all this material on the behavior of animals leave us? It appears that some animals, under the right conditions, can be taught to count up to seven. Is it really counting or is it the demonstration of another skill?

IMMEDIATE PERCEPTUAL APPREHENSION

Animals and humans have a special ability to identify a small number of similar objects presented in their visual field. Some experts call this subitizing. It is the ability to recognize a number of objects, and it does not require conscious thought. It is given to us immediately, without reflection or performance of a conscious operation. It is a kind of simple "multiplicity awareness." This is the talent being demonstrated by both the birds and the chimpanzees. For human beings, this ability is exercised on a daily basis. People, without practice, can easily distinguish one, two, three, four, and five items found in their visual field. We do not have to count; the number of items is immediately apparent. Look at Figure 5 which consists of five collections of dots from one to five. As soon as you look at each collection you know the number without counting. Now immediately say the correct number when you look at Figure 6. If you said eight, you were probably lucky. In most cases, we must count when the collection increases in number beyond five, unless the collection is ordered in a manner that directs the brain toward the right answer. This immediate awareness is subitizing. We can do it, and animals can do it too.

Human beings can actually use this immediate perception of multitude to a higher degree. On occasion, when encountering a large number of objects, we have an immediate awareness when something is wrong—when some are missing. Karl Menninger in *Number Words and Number Symbols* tells a story of South American Indians who knew only three number words, yet demonstrated the subitizing skill.[9] On a journey the Indians were accompanied by a multitude of their dogs. Upon inspection, and without taking the time to count, the Indians could immediately tell if one was missing and would call until the errant animal reappeared.

A school teacher on a field trip with her students may suddenly sense someone missing. She quickly counts the students and discovers that,

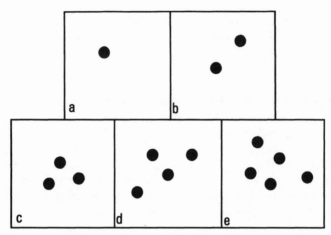

FIGURE 5. Collections of one to five dots. With each collection, the mind immediately apprehends the numbers of dots, making counting unnecessary.

yes, one is not there. Subitizing produces the first impression that something is wrong. Poker players develop a fine sense of touch for decks of cards. Frequently, after a poker hand has been played and the cards gathered in by the dealer for the next deal, he or she will announce, "We're short!" Everyone quickly checks the table before them to find the missing cards. In many cases it is only two or three cards out of fifty-two.

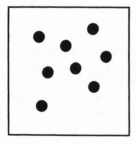

FIGURE 6. How many dots? Did you have to stop and count them?

How does the dealer know that cards are missing? He or she feels the shortage in the deck—the sensation is immediate. There are many examples of this ability to apprehend immediately magnitudes and multitudes.

However, subitizing is not counting, for there is no mapping of anything onto elements in the set. The subitizing talent in animals is good only for small numbers, below eight. When the chimpanzee sees a set of objects, it immediately recognizes the number—it does not count. When the raven sees its pattern of dots, recognition is immediate as well. There is no evidence that they consciously perform a counting operation. John McLeish in his book, *Number*, explains that an intuitive awareness is not equal to true counting.

> Animals have an intuitive awareness of number. This means that they know, from experience, without analysis and immediately, the difference between a number of objects and a smaller number. . . . This is, however, as far as it goes. Animals can only respond to a number situation when—as with eggs in the nest or food—it is connected to their species and survival needs.[10]

People and animals are born with this awareness of small numbers. This kind of awareness is necessary for the counting process to begin, but again, it is not counting, not even primitive stick-counting. So, in the sense that we have defined counting, we can conclude that we have no evidence that any animal counts at the present time. Why are we so insistent on the counting activity? You might say, "Look, the poor dumb chimp knows five dots from four dots. Give him a break—he knows small numbers even though he has no name for them." This may be true, but we are after more than just an awareness of patterns of one to seven dots. We are after the ability to perform a mapping operation. We are after an awareness of the natural numbers, which have no logical limit. We are after the ability to conceptualize not just three but thirty and three hundred. In other words, counting should be an open-ended process, and continue until it satisfies our need to determine the cardinal number of whatever set concerns us. Identifying small multitudes does not satisfy our requirements for counting.

However, our quest for animal counting is not over. While no animal at the present time is known to have the ability to count or has shown the ability to conceive of abstract numbers associated with counting, that does not exclude the possibility that some animals can be taught to count, especially the chimpanzee and dolphin. And we must not forget that whales and dolphins communicate with complex, rapid sounds, much like human beings. We have no idea whether they can count. The answer to this question awaits further research in deciphering their communications.

CAN THEY LEARN TO DO IT?

If animals do not now count, can they be taught to count at some future time? On the surface, this may seem an inane question, for if they do not count now, how could they count in the future? Still, this remains an important question. We may be tempted to say that only human beings count by mapping either words or objects onto sets. Is this because of something unique to our nervous systems, not available to other animals? If animals can be taught to count, then we will know that counting, and mathematics in general since mathematics is the study of numbers, can be made available to other species. This will demonstrate the generality and transferability of mathematics to other intelligences. If mathematics, as it is currently understood by human beings, is not transferable to other intelligent beings, then maybe mathematics, as we practice it, is not conceived in its most universal form. Maybe human mathematics reflects the uniqueness of how human beings think and does not represent a fully generalized body of truths, as claimed by many philosophers through the ages. This suggests that our mathematics would have to undergo a transformation to remove the special human viewpoint before it could be communicated to nonhuman intelligences.

The first thing we want to do as we consider animals as mathematicians is to look at the sizes of animal brains in relation to their body weights. This is a gross, imprecise measure, but it will give us an overall feel for how smart various animals might be. The assumption is that small animals have small bodies through which they receive sensations and to which they send control commands, so that the amount of nerve tissue needed is correspondingly small. Larger animals need more nerve tissue for their larger masses. Hence, if everything else were equal (which it never is) and animals shared equal intelligence, then a constant ratio should exist between brain weight and body weight. On the other hand, if an animal has more brain tissue relative to body weight than other animals, we assume that animal must be "smarter" than other animals.

Table 3 lists the brain weight in grams, the body weight in kilograms, and the ratio of body weight to brain weight for various animals. The animals are listed in decreasing order of brain size. The far right column shows the ratio of body weight to brain weight. When this ratio is small, it means the animal has lots of brain compared to body mass. Most experts recognize that this ratio is, at best, a very crude measure of intelligence. The neocortex is the outer surface of the brain's cortex and the center for higher brain functions. A better measure would be the ratio of body weight to neocortex area.

Another objection to bulk brain weight as a measure of intelligence is that birds, because they are designed by evolution to be light, tend to get more favorable ratios than they deserve. Water mammals, on the other hand, are penalized. Whales and dolphins live in the buoyant ocean and are not subject to the more severe gravity exerted on land animals. They have large stores of body fat for warmth. This tends to make them large-

TABLE 3. Ratio: Body Weight to Brain Weight

Animal	Body weight (kg)	Brain weight (g)	Ratio
Blue whale	58,000	6800	8529:1
Orca	7000	6200	1129:1
African elephant	6500	5700	1140:1
Bottle-nosed dolphin	155	1600	97:1
Human	70	1400	50:1
Common dolphin	100	840	119:1
Hippopotamus	1350	720	1875:1
Giraffe	1220	700	1743:1
Holstein cow	920	460	2000:1
Chimpanzee	52	440	118:1
Squirrel monkey	.717	26	28:1
Pygmy shrew	.0047	0.1	47:1

bodied in relation to comparable land animals. Hence, their ratios are skewed to larger numbers. Even with all these peripheral factors, we still find the body-weight-to-brain-weight ratio useful.

Table 3 shows that human beings do not have the most favorable ratio. Both the squirrel monkey and the pygmy shrew have better body weight–brain weight ratios. However, these animals have such small brains that they simply do not have the mass of neurons needed to do any serious thinking. Those animals with brains larger than the human brain, including the whale, orca, elephant, and bottle-nosed dolphin, are more interesting candidates—especially the dolphin, whose ratio approaches ours. The elephant, orca, and whale are interesting because they have such massive brains. For what do they need such huge brains? Why would a whale, which seems to occupy its time swimming around eating little shrimps, need a brain weighing about fifteen pounds? How challenging is it to swim about and open your mouth when you encounter a school of krill? Are those great leviathans using their massive brains for thinking leviathan thoughts?

Of all the animals, the most likely candidates for learning to count are probably the chimpanzees and dolphins, primarily because of the size and advanced state of their brains. As previously mentioned, chimpanzee research into counting has yielded no meaningful, positive results. When comparing brain anatomy between dolphins and human beings, the dolphins fare well (Figure 7). While major differences exist, overall the dolphin's brain is as large and complex as ours. Dolphin researchers and trainers are generally impressed with the intelligence of dolphins, and perhaps this animal is our single best bet for communicating with another species. Yet, a few claim that the dolphin's brain power is overrated. For example, David Caldwell, former associate professor at the University of Florida and head of the Division of Biocommunication, and Melba Caldwell, former research associate in the Communication Sciences Laboratory at the University of Florida, state in their book, *The World of the Bottle-Nosed Dolphin*:

> A lot of attention has been centered on the notion that dolphins are as smart as some men, or smarter than most, or that they can talk but that we humans are just too stupid to understand them. It is true that some

have worked untethered with aquanauts in the Navy's Sealab programs; however, dolphins probably are just exceptionally amiable mammals with an intelligence now considered by most workers, on a subjective basis, to be comparable to that of a better-than-average dog.[11]

Dolphins live in a world dominated by their sense of sound while we live in a world dominated by our sense of sight. When working with dolphins, we are tempted to relate to them by "showing" them objects and "showing" them hand signals. Yet, dolphins have a poorly developed sense of sight while their sense of hearing is much more advanced.

Our mathematics is also dominated by our sense of sight. We visualize our mathematical relations, instead of hearing them. Hence, even if dolphins are as smart as, or smarter than, human beings, they may never appreciate our mathematics. On the other hand, they may be able to appreciate, or even now have, an auditory mathematics. The Greek mathematician Euclid (fl. 300 B.C.) compiled a set of axioms, postulates, and theorems of geometry that were taught as the standard for over two thousand years. A basic building block of Euclid's geometry is his fifth postulate, known to millions of schoolchildren. It says: If A and B are two points, and c is a straight line passing through A (but not through B) then

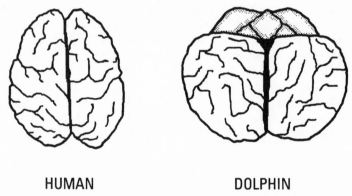

HUMAN   DOLPHIN

FIGURE 7. Drawings of typical human and dolphin brains. [Drawing adapted from Robert F. Burgess, *Secret Languages of the Sea* (New York: Dodd, Mead, & Company, 1981), p. 223.]

there exists one and only one straight line through B that is parallel to line c. We visualize this postulate in this way: First we see the two points A and B. Then we see a straight line c passing through A. Next we imagine another straight line through B that is parallel to c. The postulate says that only one line through B can be parallel to line c.

Hence we are animals that visualize. Can we imagine a comparable postulate for an animal that lives in a dim, almost black world, which is dominated by sounds? A mathematics postulate in this world might read: if A and B are tones, and c is a collection of tones in harmony with A, then there exists one and only one collection of tones harmonizing with B that are, one-to-one, an equal fraction of an octave from the tones of c (whether this sound postulate is true or even makes sense, I leave to the reader). If we were auditory animals, this kind of postulate would have a greater meaning to us than Euclid's visual postulates. What we must be sensitive to, as suggested by most dolphin researchers, is the kind of weightless, wet world the dolphins live in. We must be sensitive to the idea that sound is what makes their world real, not sight. If we can relate to dolphins in their terms, perhaps we'll find out just how smart they are, and we may be astounded.

Whales, although difficult to study, offer another opportunity for investigating large and sophisticated brains. Whales sing long, complex songs that they periodically repeat, but which they also change, implying a good memory. Individual songs may exceed half an hour in length and contain from one million to a hundred million bits of information.[12] It is difficult to ignore such impressive behavior when someone claims whales are too stupid to count.

In conclusion, we can only say that no evidence exists today that any animal counts as we define the term. However, the natures and sizes of dolphin and whale brains suggest they have the intelligence necessary for counting.

CHAPTER 4

Ancient Numbers

We have explored the history of counting from earliest hominids to modern *Homo sapiens sapiens*, the period from approximately five million years ago to eleven thousand years ago. Our story now continues with the dawn of farming in western Asia.

THE BIRTH OF FARMING

For most of our prehistory we see no evidence of writing and no mathematics beyond simple calculation, probably because humans did not have the kinds of problems more sophisticated mathematics is required to solve. Stone Age people did just fine with basic counting and possibly the rudiments of simple addition, subtraction, multiplication and division. However, once farming began, all that changed. Approximately thirteen thousand years ago people began using wooden sticks with inlaid sharp flint stones to harvest wild grains. These grains were a boon because they could be stored in the back of caves for later use. Unlike meat, grain did not spoil quickly during warm weather. Next, people began guarding the fields of wild grains and encouraging their growth through intermittent irrigation. Genuine farming occurred when people began saving some of the gathered seeds for planting the next season. This led to selecting the best seeds, and, hence, changing future crops through selective cultivation. The first major grain grown in

Western Asia was barley, which soon became the dominant currency and the source of early beer and bread.

Because hunter–gatherers relied on wild game and wild vegetation, they survived best in small groups that roamed over large areas. Farming, utilizing both domesticated animals and crops, provided for much denser populations than the hunting–gathering lifestyle. Soon villages grew into farming centers. While this occurred independently in several locations on the globe, the earliest such farming occurred around 11,000 B.C. in the Fertile Crescent, that area stretching from ancient Jericho of Israel's West Bank, north to Damascus and Aleppo (now Haleb) in Syria, and then southeast to Baghdad and Basra of Iraq and even to Susa, an ancient site in Iran (Figure 8). The first of these farming villages appeared in southern Turkey and northern Iraq in the Zagros hills. Later villages include Jericho, which flourished around 8000 B.C. With the growth of farming villages other crafts soon emerged. Pottery-making began about 6500 B.C. while weaving and wheeled wagons appeared about 6000 B.C.

What kinds of problems did farmers living in villages have to solve that their cousins, the hunter–gatherers, never confronted? A primary advantage of farming is that it can generate food surpluses. It is this surplus of grain that allows a population to stay in one place and construct permanent homes, but only if this grain can be protected and its volume measured. Some grain can be used during the current year for food, but a portion must be reserved for next year's seed, and an additional portion may be needed to trade for goods transported from other farming villages. In addition, the cropland itself must be partitioned properly and protected. A new dimension to numeration is born with land use, for not only must distinct items be counted, but land must be measured. Therefore, the association between natural numbers and measurement was established.

In addition to managing the fields and crops, the greater numbers of people must be efficiently organized to plant, tend, and harvest the crop; and then they must be armed to protect the crop and land against raiders. These activities require more than just counting. Large numbers of people, food stores, and plots of land put pressure on the scribes and priests to add, multiply, subtract, and divide. A ruling class evolved whose scribes computed taxes on individuals and households.

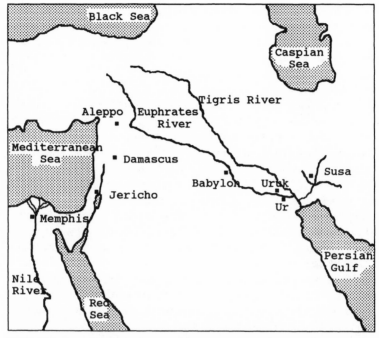

FIGURE 8. Map of the Fertile Crescent area of western Asia. The Fertile Crescent begins with Jericho in the west, moves north to Aleppo, and then southeast down the Tigris and Euphrates river valley.

The season of the year and the weather conditions took on added importance once farming took hold. Farmers needed a calendar to identify the proper time to plant each crop. This required more computing skills to keep track of the positions of the stars and project those positions forward in time. An example of calendar-calculating prior to the invention of writing (3500–3100 B.C.) is the ancient Babylonian year, which began with the vernal equinox (March 21). The first month was the month of Taurus, the bull. This suggests that the calendar was established at a time when the sun was in Taurus around March 21. Such a period began about 4700 B.C. The Sumerians, an even more ancient people, may have established their calendar around 5700 B.C.[1]

All these necessary activities—accounting for food stores, land, people, and seasons—stimulated the rulers and accountants to invent faster, easier ways of manipulating the natural numbers. Three different solutions to keeping numerical records evolved: tally sticks, knotted ropes, and clay tokens.

Tally sticks are bones or wooden sticks that have been deliberately marked or etched to indicate numbers. The previously mentioned thirty-thousand-year-old wolf bone is an example of a primitive tally stick. Another example is a bone discovered at a fishing site on Lake Edward in Zaire that dates between 9000 and 6500 B.C. Both these bones have numerous notches cut into number patterns. Most tally sticks, however, are wooden. They have been used for millennia by ordinary people to keep track of goods, transactions, and contracts (Figure 9). A popular form was the double tally stick used to keep track of debts. A long thin stick was cut with notches to indicate the size of the debt. It was then split for almost the entire length, and then separated into two matching sticks. The stick with the large end was the stock, and was retained by the lender. The stick with the small end was the inset, and given to the debtor. At a later date, when the two parts of the stick were rejoined, both the lender and debtor could tell if any tampering had been done to the notches, for the two parts should match perfectly.

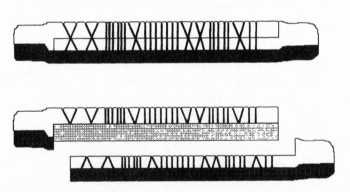

FIGURE 9. Drawing of a Finnish tally stick recording work completed. The length is 25 cm. [Drawing adapted from photograph, Karl Menninger, *Number Words and Number Symbols* (New York: Dover Publications, 1969), p. 231.]

Tally sticks were used across the globe in Africa, Europe, the Pacific Islands, America, and China. In Latin, tally (*talea*) means "cut twig." The Chinese written symbol for contract has three parts: on top are the two characters for a notched stick and a knife while below is the character for large. Hence, the symbol stands for "large notched stick"—tally stick. An interesting historical note on tally sticks comes from Britain, where tally sticks were used until 1828 for tax records. Large piles of these tally sticks were stored in the Houses of Parliament. In 1834 the British Government decided to destroy them by burning, with the result that the Houses of Parliament were accidentally burned to the ground.

A second physical record for numbers is the knotted cord or rope. This method was also widespread, found in such diverse locations as Africa, North and South America, and China. Lao-tse, the fifth-century-B.C. Chinese philosopher, recommended that people return to tying knots in cords as a method of writing. The most advanced knotted cords were produced by the Incas of Peru. They called their cords *quipus* and used them to record official transactions (Figure 10). The quipus contained three different kinds of knots and also varied in length, color, and the positioning of the knotted clusters. The method was so sophisticated that even nonnumerical information was recorded on them.

The third physical method of recording numbers is the most significant: clay tokens. For many decades archaeological researchers were baffled by small clay artifacts found at Neolithic (8000 to 3500 B.C.) city sites in the Fertile Crescent. These artifacts or tokens were uncovered in most early farming sites from Chanhu Daro in Pakistan, to Beldibi in southwestern Turkey and south to Khartoum in the Sudan. The oldest tokens come from sites in Iran and go back to 8000 B.C. The tokens take the forms of spheres, disks, cones, ovoids, triangles, rectangles, and other assorted odd shapes. Generally, they were from one to four centimeters (0.4 to 1.6 inches) in length (Figure 11).

Archaeologists invented numerous functions for these tokens including children's toys, pieces to games, phallic symbols, female figurines, nails, and marbles. However, if the theory originated by Denise Schmandt-Besserat of the University of Texas is correct, then none of these guesses is the right one. During the 1970s and 1980s, Schmandt-Besserat conducted extensive research on ten thousand tokens from 116

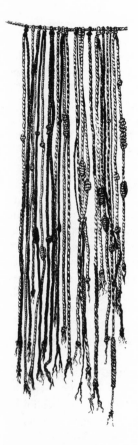

FIGURE 10. Inca quipu rope. The knots are digits. Numbers are read from the top of a string to the bottom. (Photograph from Stock Montage, Chicago, IL.)

sites, and discovered that they represented specific objects and were used for record keeping.[2] In her book, *Before Writing*, Schmandt-Besserat says:

> Tokens can be traced to the Neolithic period, starting about 8000 B.C. They evolved following the needs of the economy, at first keeping track of the products of farming and expanding in the urban age to keep track of products manufactured in workshops.[3]

According to this thesis, tokens were in extensive use for accounting from the beginning of the eighth millennium (five thousand years before

FIGURE 11. Counting tokens from western Asia. (Photograph courtesy of Denise Schmandt-Besserat and Musée du Louvre, Paris, France.)

the invention of writing). The original tokens used from 8000 to 4400 B.C. were simple in shape and design with different shapes representing different objects. For example, an ovoid represented a jar of oil while a small sphere was one measure of grain.[4] After 4400 B.C., many more token designs were used and the markings on the tokens became more complex and numerous. During this early farming period (8000–3100 B.C.), the tokens were used in a one-to-one relationship to the items accounted for. Three jars of oil would be counted with three ovoids, and four measures of grain would be counted with four spheres. This was an advanced kind of stick-counting, where the tokens were fashioned to identify the items counted. At this time tokens were not used to represent abstract numbers.

THE SUMERIANS INVENT WRITING

During the Chalcolithic Age (4500–3000 B.C.), a startling change occurred in the Fertile Crescent that was almost as dramatic as the beginning of farming. The change was the growth of true cities from small farming villages. The growth of cities coincided with the use of the more complex tokens. The first cities appeared in the lower part of the Mesopotamian plain in an area known as Sumer, which is now southern Iraq. For five millennia remnants of this remarkable civilization lay hidden under the sands of the Iraqi desert, and its existence was not even imagined until discovered during the first half of the nineteenth century. The Sumerians arrived in southern Iraq about 3500 B.C. and built an empire that lasted until the Babylonians conquered it around 2000 B.C. By 3000 B.C. more than a dozen cities covered the Sumer area with Ur being the largest. It contained approximately twenty-five thousand individuals with as many as two hundred thousand in the surrounding farming villages. By this time other cities had sprung up in the Fertile Crescent.[5]

The new cities of western Asia generated a surplus of manpower whose labors were directed toward specialized trades and the manufacture of goods. For the cities to flourish, goods and raw materials had to be traded with other cities. For example, the cities of Sumer had no local access to good timber or stone, or to metals such as copper, silver, and gold. The need to transport goods and verify the size of shipments made greater demands on the accountants. It was the Sumerians' attempt to solve these problems that resulted in complex tokens and eventually led to writing.

When a shipment of barley, livestock, or manufactured goods was to be transported to a new location, a record had to accompany the shipment so the purchaser would know that he or she had not been cheated by either the seller or those responsible for transportation. The record consisted of a collection of tokens that specified the number and type of goods being shipped. But the seller could not just place these tokens in a bag since anyone could steal goods and then steal the appropriate tokens to make the shipment appear complete. Therefore, the Fertile Crescent cities devised a clever technique to protect the tokens. They wrapped the tokens in clay and then baked the clay. This provided a hard clay ball with the clay tokens inside. When it was delivered to the buyer, the buyer could break it open and verify that the tokens inside agreed with the delivered goods. Ingenious!

These token-filled balls of clay are called *bullae* or *envelopes* and were first found at the ancient site of Susa in Iran. The oldest envelopes come from the site of Farukhabad, approximately 150 kilometers north of Susa, and date from 3700–3500 B.C.[6] However, the envelopes discovered at archaeological sites included an additional feature. If you were a shipper and were responsible for transferring goods to a second shipper who would then deliver the goods, you had to demonstrate that you were handing over the correct amounts. How could you do this when the clay tokens were hidden inside the envelope? The answer was to imprint or draw a picture of the tokens contained inside the envelope on its outside surface before it was baked. Anyone in possession of the envelope could then simply "read" on the outside which tokens were contained inside. Hence, when you gave a shipper the shipment with its envelope, he could read on its outside surface what was supposed to be in the shipment. At the final destination, the buyer could open the envelope and verify both that the markings on the outside agreed with the tokens inside and that the shipment was complete.

The above system worked well until the clever Sumerians realized that the tokens contained within the envelopes were superfluous. All that was really needed was the imprinted record of the tokens baked into a clay surface. Hence, they abandoned the envelopes for simple clay tablets, and writing was born. At first, the clay tablets retained the basic characteristics of the imprinted envelopes. Each token still represented one specific item, for numbers had not been abstracted from the objects counted. This type of accounting is called *concrete counting* by Denise Schmandt-Besserat, and was used on envelopes and tablets from 3500–3100 B.C. Finally, around 3100 B.C., the Sumerians separated the impressions representing the number of items from the items themselves. The abstract numbers were impressed on the clay while a pictogram of the object being counted was inscribed with a writing implement upon the clay (Figure 13). This separation of numbers and objects was much more efficient, facilitating the use of cardinal numbers to indicate quantity in conjunction with a single pictogram to identify the item. Once items were no longer associated with numbers, the pictograms representing them could be generalized to represent many different ideas, and writing was born. Schmandt-Besserat summarizes for us:

> The tokens give new insights into the nature of writing. They establish that in the Near East writing emerged from a counting device and that, in fact, writing was the by-product of abstract counting. When the concepts of numbers and that of items counted were abstracted, the pictographs were no longer confined to indicating numbers of units of goods in one-to-one correspondence. With the invention of numerals, pictography was no longer restricted to accounting but could open to other fields of human endeavor. . . . The invention of abstract numerals was the beginning of mathematics, and it was also the beginning of writing.[7]

One of the most remarkable creations of the human race, writing, was achieved by the Sumerians between 3500–3100 B.C. Prior to this time, the legacies of cultures had to be handed down to succeeding generations by word of mouth and by demonstration. With writing, all that changed. People not only could pass information on to their own immediate successors, but also could pass it to distant peoples and distant generations. It is by examining the first writings that we get a glimpse of how people lived and how they calculated in ancient times. And it was the need to record numbers that first gave the impetus to the development of writing.

The earliest Sumerian writing was simply the impression of tokens into soft clay around 3500 B.C. This was followed by the impression of tokens to represent numbers plus the scripting of pictograms to represent objects around 3100 B.C. To inscribe on the clay, a wooden or bone stylus in the shape of a cylindrical pencil was used. The purpose of such writing was to record dates and keep lists of objects. Around 3000 B.C. the Sumerians developed cuneiform writing from the earlier writing. With cuneiform, the token ideograms of number-graphic writing were replaced with phonetic symbols, and a separate abstract numbering system evolved. The earlier writing was in columns from right to left. This changed after 3000 B.C. to writing on lines from left to right. Fortunately for us, the writing medium was baked clay, which preserved the efforts of the Sumerians for over five thousand years.

SUMERIAN MATHEMATICS

Little is known of Sumerian mathematics, but it is obvious from their surviving tablets that they knew the four basic operations of arithmetic: addition, subtraction, multiplication, and division. The complexity of the Sumerian society required skill in manipulating the natural numbers. From the first tokens for numbers we see that the Sumerian numbering system was sophisticated compared to the earlier 2-counting, 5-counting, and even 10-counting systems, for it was a sexagesimal system based on both sixty and ten. Many of the tablets appear to be the work of student scribes practicing their lessons. From these tablets we know that the Sumerians could work with both very large numbers and very small numbers, using both whole numbers and fractions. To inscribe the number "one," the scribe would press a cylindrical stylus at an angle into the clay. This left a semicircle which looked something like a capital "D" resting on its side (Figure 12). The symbol for "one" was repeated to make the larger numbers. A collection of numerals that represented a number were separated from other writing by placing them in a rectangle. When ten was reached, the stylus was pushed vertically into the clay,

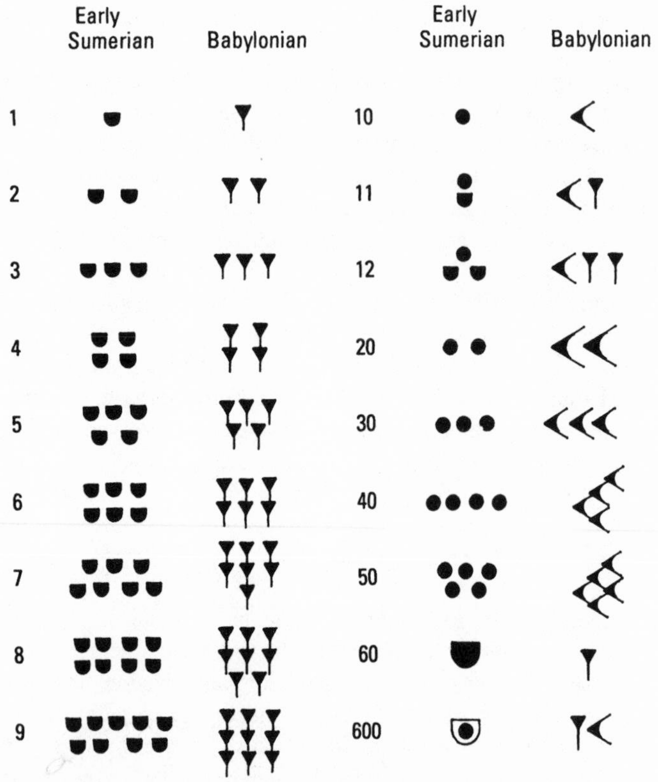

	Early Sumerian	Babylonian		Early Sumerian	Babylonian
1			10		
2			11		
3			12		
4			20		
5			30		
6			40		
7			50		
8			60		
9			600		

FIGURE 12. Early Sumerian and Babylonian numbers. Sumerian cuneiform writing was adopted by the Babylonians, hence the later Sumerian numbers were almost identical to Babylonian numbers.

making a small circle (○). But the Sumerian system was not based on ten, but on sixty. From one to fifty-nine a combination of ones (◗) and tens (○) were used. For sixty a large D-shaped symbol was drawn into the clay. The next step was 600 or 60·10, which was a large D with a small circle inside (◉). 60·60 or 3,600 was a large drawn circle (○) and 36,000 (10·60·60) was a large circle with a small circle inside (◎). The various forms for these early period Sumerian numbers are given in Figure 12.

Our modern system depends on the position of each digit in the number. Hence, thirty-seven is different than seventy-three because the two digits, three and seven, occupy different positions. Hence, ours is a positional number system. The early Sumerian numbering system was a nonpositional system since the scribe simply added the values of the various symbols to arrive at the intended sum. There was no place value on the symbols as in our modern system. In later cuneiform writing, the end of the stylus was shaped so that it made triangular marks with tails in the clay.

By about 2400 B.C. we see that the Sumerians were drafting checks, measuring land in *shars*, weighing in talents (*gur*), measuring liquids in *ka*, and computing interest. Most remarkably, they were using the fractions one-half, one-third, and five-sixths. This is the earliest known recognition that fractions are numbers. The outstanding characteristics of the Sumerian system are that it was based on sixty with ten as an intermediate step and that fractions had begun to appear. This allowed the Sumerians to write not only large numbers but small numbers as well.

FIGURE 13. Sumerian clay tablet. Impressed in the clay are the numerals for 33. Drawn into the clay is a pictogram of a jar of oil. These early numerals for 1 have a long, drawn-out appearance. (Courtesy of Denise Schmandt-Besserat, of the University of Texas at Austin, and the Royal Ontario Museum, Ontario, Canada.)

We cannot emphasize the importance of Sumerian fractions too much. Since the beginning of named number counting, when numbers began to be abstracted, the only numbers available were the natural numbers. Now a new kind of number was introduced. If the first numbers were given names back with the evolution of rapid phonetic speech, approximately one hundred thousand years ago, then the journey from natural numbers to fractions took approximately one hundred millennia! Fractions were a long time coming. In fact, we will see that both the Egyptians and the Greeks did not embrace the idea of fractions as fully as the Sumerians.

THE REMARKABLE BABYLONIANS

About 2000 B.C. a people known as the Amorites invaded Sumer and conquered its cities, destroying Ur in the year 2006 B.C. These people became known as the Babylonians and forged a large empire that included areas of Iraq, Jordan, and Syria. They built numerous cities, including the famous Babylon, and their empire lasted until 538 B.C., when Babylon fell to Cyrus of Persia.

After the Babylonians conquered Sumer, they adopted both the Sumerian cuneiform writing and the Sumerian mathematics. Thousands of Babylonian cuneiform tablets have been found, especially from the periods of 2000 B.C. and 600 B.C. The older records have yielded a wealth of information on Babylonian mathematics. While the Babylonians retained the basic Sumerian numbering system based on ten and sixty, they abandoned the special symbols for 60, 10·60, 60^2, 10·60^2, and 60^3 and retained only two symbols: the triangle with a vertical tail, called a wedge ($\mathbb{T}$), that stood for 1; and a triangle with two side tails, called a hook ($<$), for 10. (Three exceptions to this are the fractions one-half, one-third, and two-thirds, each of which had its own special symbol.) Then they added a feature that made their numbering system almost unique among the ancient peoples. They used a positional system which could represent both large and small numbers.

Our current decimal system is a positional system that allows us to represent both large and small numbers and to carry out complex calculations. For example, 1 stands for just the number "one," 10 is the number "ten," while 100 is the number "one hundred." The position of the "1" indicates the numerical value to be attached to it. Consider the number 743. In a nonpositional system this number would be $7 + 4 + 3 = 14$. But in our system it is really $7 \cdot 100 + 4 \cdot 10 + 3 = 700 + 40 + 3$ or seven hundred and forty-three. If we want to show fractions with our system we place a "decimal point" on the right and the next digits represent 10ths, 100ths, etc. Hence, 57.32 is $5 \cdot 10 + 7 + 3 \cdot (\frac{1}{10}) + 2 \cdot (\frac{1}{100}) = 50 + 7 + 0.3 + 0.02 = 57.32$.

The Babylonian positional system worked the same way but was based on sixty, not ten. Therefore, the number (here written in Arabic digits) of 621 is not six hundred and twenty-one but:

$$6 \cdot 60^2 + 2 \cdot 60 + 1 = 21,600 + 120 + 1 = 21,721$$

Of course, when writing the symbol for 621 they did not use our Arabic symbols, but their own triangular wedges. Hence, the number for 21,721 became:

𒌋𒌋𒌋 𒐏 𒐕

This positional system allowed them to show fractions. For example, 4.5 would be 𒐏 followed on the right by ⪡⪡⪡. The 𒐏 gave them the 4 in 4.5 and the ⪡⪡⪡ (or thirty) gave them 30/60ths which is the same as 5/10ths. Hence, with only the symbols for one (𒐕) and ten (⪡), the Babylonians, using their positional system, could write very large and very small numbers and could use them to efficiently carry out calculations. Scholars have wondered why the base of sixty was chosen by the Sumerians and used by the Babylonians. (It probably evolved prior to the Sumerians as part of the token system used in preliterate western Asia.) One possible explanation is that sixty can be evenly divided by many smaller numbers. That is, sixty is evenly divisible by two, three, four, five, six, ten, twelve, fifteen, and thirty. This makes many basic calculations simple. The Sumerian and Babylonian systems were extensively used to compute weights, measures, and land areas—all of which would be facilitated by the divisibility of sixty.

Unfortunately, the Babylonian system suffered from two significant defects. First, there was no placeholder, such as our zero, to show that one of the powers of sixty was empty. Second, there was no decimal point to show where the fractional part of the number began. The reader was supposed to figure this out from the context of the problem. This made Babylonian written numbers ambiguous. For example, the number ⟨Π Π ⟨⟨⟨ could be any of the following:

(a) $12 \cdot 60^2 + 2 \cdot 60 + 30 = 43,350$
(b) $12 \cdot 60^3 + 2 + 30 \cdot (\frac{1}{60}) = 216,002.5$
(c) $12 \cdot 60 + 2 + 30 \cdot (\frac{1}{60}) = 722.5$
(d) $12 + 2 \cdot (\frac{1}{60}) + 30 \cdot (\frac{1}{60})^2 = 12.041666$ (approximately)

From the context of the problem being solved, the scribe was supposed to realize how the values of twelve, two, and thirty were being used in the positional system. A position holder or zero was not in use until the time of Alexander the Great, two centuries after the end of the Babylonian empire. This was achieved by a special symbol of two oblique wedges inserted between numerals to act as a zero.

Even with the omission of a zero and a decimal point, the Babylonian numbering system was superior to both the Egyptian and Greek nonpositional systems. The Babylonians clearly achieved a high degree of mathematical sophistication. Many of the tablets unearthed are mathematical tables for multiplication, computing squares of numbers, reciprocals, cubes, square roots, and cube roots, and tables for computing interest. The Babylonians also knew how to use algebra to solve problems well beyond the level reached by the Egyptians. They knew how to add and multiply both sides of an equation to simplify it. They could carry out simple factoring of terms. They did not use letters to represent quantities as we do, but used terms such as volume, breadth, and length.[8]

The Babylonians could even solve simultaneous equations in two unknowns, certain quadratic equations, and some cubic equations. They were even aware of the Pythagorean theorem, which says that the sum of the squares of the two legs of a right triangle is equal to the square of the diagonal side, which is symbolically written as $A^2 + B^2 = C^2$ where A and B are the lengths of two sides and C is the length of the diagonal (Figure 14). Ancient scholars gave credit for its discovery to the Greek mathematician Pythagoras who lived in the sixth century B.C. However, later discoveries have made it apparent that both the Babylonians and the

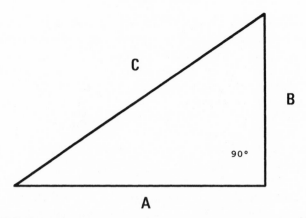

C

B

90°

A

FIGURE 14. Right triangle demonstrating the Pythagorean theorem: $A^2 + B^2 = C^2$.

Chinese were aware of this theorem long before Pythagoras. There exist an infinite number of whole number triplets that satisfy this theorem. Such number triplets are called Pythagorean numbers. For example, the whole number triplets three, four, and five satisfy the Pythagorean theorem, for if we have a right triangle whose legs are of length three and four, then the longer diagonal side will be of length five. We see also that

$$3^2 + 4^2 = 5^2 \text{ or } 9 + 16 = 25$$

The Babylonians knew of Pythagorean numbers since they constructed tables of them. Yet, they seem to have gone even one step further. If we have a right triangle with both legs of length 1, then the diagonal has length $\sqrt{2}$. We can see this by letting X stand for the length of the diagonal and then solving for X using the Pythagorean theorem.

$$X^2 = 1^2 + 1^2 = 1 + 1 = 2$$

If X^2 equals 2 then X equals $\sqrt{2}$. But what is this number, $\sqrt{2}$? As it turns out, it is neither a natural number nor a fraction, but an entirely new kind of number. In one of their tablets, the Babylonians actually computed $\sqrt{2}$ to be 1.4142129. An approximation of $\sqrt{2}$ carried out to ten places is 1.414213562. Hence, the Babylonian estimate is off by only about 0.0000007, a very small amount indeed.[9] Did the Babylonians realize they were dealing with a new kind of number? There is no evidence that they did, yet it is intriguing that they came so close to discovering *irrational* numbers, numbers that had to wait some fifteen more centuries to be identified.

The Babylonians applied mathematics to the science of astronomy. They were able to measure the periodic motions of the moon, sun, and five visible planets: Mercury, Venus, Mars, Jupiter, and Saturn. From these measurements they constructed an arithmetic model to predict future conjunctions and alignments against the background of the stars.

So, what was missing from Babylonian mathematics? We have already dealt with the lack of a zero and a decimal point in their numbering system. In addition, their use of mathematics seems to have been directed toward solving specific problems and not illustrating general laws. They did not distinguish between exact and approximate solutions and they appear to have had no need for mathematical proofs. Yet, even with these deficiencies, the Babylonians achieved the highest degree of mathematical competency of their time, surpassing their contemporaries, the Egyptians.

To summarize the achievements of the people of western Asia as demonstrated by the Sumerians and Babylonians, we can list the following significant highlights:

(a) Writing was invented.
(b) Fractions were introduced.
(c) A positional numbering system was developed.
(d) Mathematics evolved from the simple counting of things (accounting) to problem solving (algebra) and measuring magnitudes (geometry).

Tens of thousands of Sumerian and Babylonian clay tablets have been found and many of them are still undeciphered. Since major ancient cities have been unearthed as late as the 1970s (e.g., Ebla of Syria), we can see that our history of this time period for western Asia is still being written. New discoveries may require us to reassess the mathematical achievements of these remarkable people.

THE EGYPTIANS

Because they accomplished such great feats, and because their empire lasted for such a long time, historians frequently give great credit to the Egyptians for their mathematics. Careful research shows such credit to be somewhat exaggerated.

Ancient Egypt was a marvel by most standards. Their civilization probably began sometime in the fifth millennium B.C. and lasted until they were conquered by Alexander the Great in 332 B.C., a period of roughly four thousand years. The early Egyptians lived in two kingdoms: Upper Egypt in the Nile River Valley, and Lower Egypt on the Nile River Delta. They became a great empire when Menes united these two halves and established his capital at Memphis sometime between 3500 and 3000 B.C. The Egyptians were a great world power for the next three thousand years. The only disruption came between 1720 B.C. and 1570 B.C. when the Hyksos from Asia Minor conquered the Egyptian delta.

The Egyptians reached the pinnacle of their early civilization from the thirtieth century until their third dynasty, about 2500 B.C. During this period they built the great pyramids; and, even though no direct evidence exists, it was probably during this time that they developed the basis for their mathematics. Their ascent to greatness was so rapid that just before 3000 B.C. they were learning to cut stone masonry for buildings, and by the beginning of the twenty-ninth century they were building the Great Pyramid of Giza.[10]

The strength of the Egyptian civilization, based on the abundance of wheat grown along the Nile river, provided a permanence and consistency for the Egyptians. This strength is reflected in their great monuments and the phenomenal longevity of their society, yet it is also reflected in the fact that their mathematics evolved little, if at all, from the third millennium on. There was no need for better mathematics since the old was entirely adequate.

Two kinds of writing were developed by the Egyptians. The first writing was the hieroglyphics, which started as simple pictorial signs around 3000 B.C., and may have been influenced by the cuneiform of the Sumerians. Hieroglyphics evolved into a combination of pictograms and phonetic signs, and were used for formal writing on monuments and temples. It was primarily meant for recording the accomplishments of individual pharaohs, and lasted until the first century B.C.

The second style of writing is called hieratic, and was used for the day-to-day management of the empire. Hieratic was less formal than hieroglyphics and more abstract, generally requiring fewer brush strokes. It is through hieratic writing that we have learned the most about Egyptian mathematics. Thousands of documents written in ink on papyri scrolls have been preserved, but most date from the end of the empire. The vast preponderance of our knowledge of Egyptian mathematics comes from two famous papyri scrolls, the Rhind papyrus, dated approximately 1650 B.C. but evidently copied from an earlier scroll that was probably written between 2000 and 1800 B.C., and the Moscow papyrus, dated around 1890 B.C. Both papyri contain examples of common practical problems facing Egyptian scribes. The Rhind papyrus is the more extensive document with eighty-four problems while the Moscow papyrus contains twenty-five. Scribes, responsible for writing on papyrus in hieratic, were of a different class from the priests, usually being trusted slaves. They were engaged as secretaries and accountants, working in the temples, which served as government offices.

The Egyptian number system was based on ten. Hieroglyphics used a combination of strokes (❘) to represent numbers from one through nine while the cursive hieratic used strokes for one through three but unique symbols for the numbers four through nine (Figure 15). Additional symbols represented the numbers ten, one hundred, one thousand, ten thousand, and up. There was no value assigned to the position of a symbol, so the system was strictly additive. To read a number, one simply added the values for the different signs. This can be cumbersome for large numbers, yet the Egyptians did manage to record numbers into the millions. Early writings frequently reflected the personality of the author with his or her own particular choice for many symbols. It was not until the fifteenth century B.C. that the writing became reasonably standardized.

Like the Babylonians, the Egyptians used fractions. However, their system was much more awkward. With the exception of two-thirds and three-fourths, all of their fractions were unit fractions or fractions where the numerator was one while the denominator was any whole number, hence all fractions were of the form 1/n. This meant that any proper

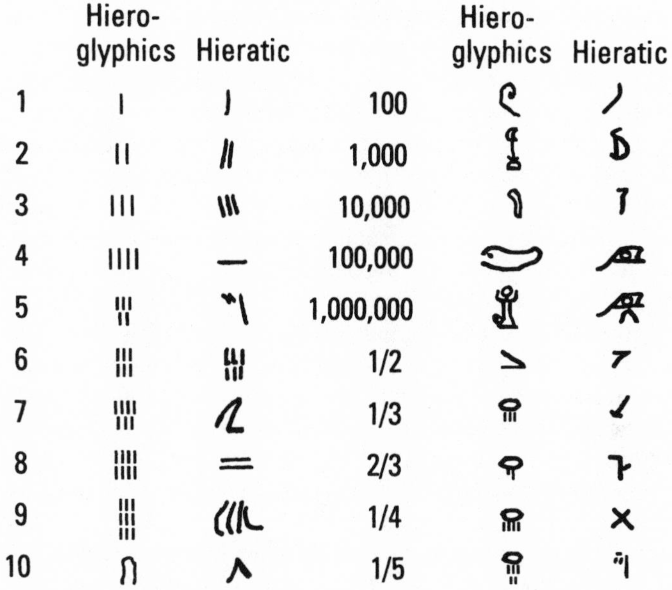

	Hieroglyphics	Hieratic		Hieroglyphics	Hieratic
1	I	J	100	ℂ	⟩
2	II	//	1,000	🯅	𝔇
3	III	\\\\	10,000	𝌩	𝟕
4	IIII	—	100,000	⟅	⬀
5	⁝⁝	⟍	1,000,000	𝕏	⟁
6	⁝⁝⁝	𝔲	1/2	⟍	�7
7	⁝⁝⁝⁝	𝓁	1/3	⟁	✓
8	⁝⁝⁝⁝	=	2/3	⟊	ⴕ
9	⁝⁝⁝	⟪⟪𝓁	1/4	⟊	×
10	∩	⋀	1/5	𝌩	⁇|

FIGURE 15. Ancient Egyptian hieroglyphics and hieratic numerals.

fraction with a numerator larger than one had to be represented as the sum of unit fractions. For example, $\frac{2}{7}$ might have been written as $\frac{1}{4} + \frac{1}{28}$. Other fractions would have required a more complex set of unit fractions; for example, $\frac{13}{21}$ became $\frac{1}{2} + \frac{1}{9} + \frac{1}{126}$. It is immediately apparent that representing all fractions as unit fractions was a messy and cumbersome requirement. To carry out computations with unit fractions, the Egyptians had to resort to extensive use of tables. For example, one table showed the combination of unit fractions resulting from doubling unit fractions, that is, the unit fractions for fractions of the form 2/n. There is no unique way to write a fraction as a sum of unit fractions. This means two different scribes could represent the same fraction in two ways, making comparisons difficult.

The extensive use of unit fractions may have come about from the Egyptian notation system. In hieroglyphics, the fraction was written as an oval above the whole number (Figure 15) while in hieratic, beginning with one-fifth, the fraction was frequently written as a dot above the whole number. Hence, the notational system made it simple to write unit fractions, since only the denominator needed to be written, and more difficult to write any fractions other than unit fractions. Special symbols were used for one-half, two-thirds, and one-fourth, but the Egyptian commitment to the unit fraction stifled their development of numbers. The impact of the Egyptian use of unit fractions extended far beyond their own civilization, and unit fractions were used by the Romans into Europe's medieval period.[11]

Even though the Egyptian handling of fractions was generally inferior to the Babylonian sexagesimal system based on sixty, it did have the advantage of showing which numbers were fractions, while the Babylonian system required the reader to guess which numbers were fractions from the context. The Egyptians were deft at using all four of the operations of arithmetic, but the calculations for both multiplication and division were additive in nature. To multiply two numbers, one simply listed the multiples of one of the numbers beginning with one and advancing in powers of two to two, four, eight, and so on, as in the following example of multiplying 7 times 11:

$$
\begin{array}{rr}
/\ 1 & 7 \\
/\ 2 & 14 \\
4 & 28 \\
/\ 8 & 56 \\
\end{array}
$$

A hash mark has been placed next to those multiples of seven that add up to eleven, namely $1 + 2 + 8$. We add the corresponding numbers in the right column for our result: $7 + 14 + 56 = 77$. This multiplication by doubling was used for centuries after the Egyptians, and, because of its popularity among the Eastern Europeans, is now called the Russian peasant method. It has been a favorite method of ordinary people to do multiplication problems encountered in daily life.

Division is a little trickier, but is based on the same general procedure. If we want to divide 187 by 11 we begin by multiplying and halving the divisor until the correct combination is reached to add to the dividend.

$$/ \quad 1 \qquad 11$$
$$2 \qquad 22$$
$$4 \qquad 44$$
$$8 \qquad 88$$
$$/ \ 16 \qquad 176$$

From the right column we see that $11 + 176 = 187$. Therefore, we add the corresponding numbers in the left column to arrive at our quotient or $1 + 16 = 17$.

In addition to the four operations of arithmetic, the Egyptians could do some limited algebra and geometry in the form of story problems, since they had almost no symbols to represent algebraic operations. The sign for addition and subtraction was the symbol of a man walking into (addition) or out of (subtraction) a house. In algebra they solved linear equations in one unknown of the form $x + ax = b$ and $x + ax + bx = c$. Some simple quadratic equations in one or two unknowns were solved. In modern notation, examples of such equations are: $ax^2 + bx + c = 0$ or $ax^2 + by^2 = c$. In geometry they had sets of rules for computing areas and volumes, some of them correct and some leading only to approximations. In solving many of their algebraic problems they resorted to a method called false position. In this method, the scribe guessed at the correct value for the unknown, called a heap, in the equation. He then substituted this value in the equation to see if it worked. If it did not, he adjusted it proportionally until he obtained the correct answer.

Although the Egyptians are often given credit for knowing the Pythagorean theorem, there is no direct evidence of this from their writings. However, their surveyors probably did know that a rope divided into lengths of three, four, and five units would, if formed into a triangle, result in a right triangle. Like the Babylonians, the Egyptians were unaware of irrational numbers (numbers that cannot be represented as the ratio between two integers p/q) and simply used whole numbers and fractions to represent the square roots of numbers that occurred in algebraic problems. One of their estimates for π was based on their formula for the area of a circle, which was $A = (8D/9)^2$ where A is the area and D is the diameter. This implied a value for π of 3.1605 which is less than 1% in error.

Much has been made of the dimensions of the pyramids, with claims that they reveal that Egyptians knew certain advanced mathematical relationships. The evidence from their writings does not support this assertion. Even though Egyptian mathematics was behind that of the Babylonians, the Egyptians did accomplish some amazing feats of measurement. According to David E. Smith in his *History of Mathematics*, the maximum error in the length of the sides of the Great Pyramid is only 0.63 inches, which, he points out, is only 1/14,000 of the total length. A second example of accuracy is that the error in the angle at the pyramid's corners is no more than twelve seconds of a degree, which is a mere 1/27,000 of the ninety degrees in a right angle.[12] How much of this accuracy is due to careful craftsmanship and how much is based on sophisticated mathematics is unknown.

Beside the everyday problems of dividing quantities of bread and beer for workers, and measuring for buildings, the Egyptians needed a good calendar. They relied on the wheat grown in the Nile Valley and had to track the river's flooding and its changing levels as a function of time. The Egyptian year was twelve thirty-day months plus five feast days for a total of 365 days. This was off by a quarter of a day and was not corrected by adding one extra day in February every four years as we do. This meant that the seasons slowly shifted out of alignment with the calendar. After 1,460 years, the days would once again match the seasons. Using this fact, it has been suggested that the Egyptian calendar was adopted either in 4241 B.C. or 2773 B.C.[13] However, the Egyptians did not base their calendar on entire star constellations, as did the Babylonians, but on the rising of the single star, Sirius. Therefore, their astronomy was inferior to Babylonian astronomy. Their calendar began on the first day in the summer that Sirius became visible above the horizon just before sunrise. This day was probably chosen because it was the day the Nile river waters began to rise.

In summary, we can say that Egyptian mathematics was very old, probably evolving during the first half of the third millennia before Christ, and then becoming institutionalized with little additional development.[14] Egyptian mathematics was inferior to that of the Babylonians, remaining almost static for two to three thousand years. In relation to numbers, the Egyptians had a more primitive notion of fractions than the

Babylonians, failing to realize that quantities could be represented by a single fraction rather than a combination of unit fractions. They used mathematics as a practical tool, and nowhere do we encounter mathematical proofs. From the kinds of problems presented in the papyri, we can guess that they sometimes solved problems for recreational purposes.[15]

Why did Egyptian mathematics stagnate after such an auspicious start? If we were to assume that mathematics depends on the free time available to priests and scribes, then we would conclude that Egyptian mathematics should have continued to evolve to a high degree during the thousands of years of the Egyptian empire. As we pointed out earlier, mathematics and science do not develop from idle time, but out of necessity. When the Egyptian society was organizing and building pyramids from 3500 to 2500 B.C., they needed a mathematics to solve everyday practical problems. Once this was accomplished, their mathematics became institutionalized and ceased to evolve. Surely there is a lesson here for us.

NUMBERS, CALCULATING, AND PROBLEMS

We have seen that basic counting answers the question: how many? The ability to compute the cardinal number of a set solves the problem of how to account for the elements of the set. Early human beings needed a method to account for the exact manyness of collections of items, and the discovery of natural numbers and counting was the solution to that problem. As we continue in our exploration of numbers, we will see that the discovery of mathematics, in general, is a response to solving problems. This realization will help us to acquire a better understanding of what mathematics is, and how we, as human beings, relate to it.

Hunter–gatherers counted, and this level of sophistication seemed adequate. When farming began, a host of new problems faced our ancestors as they tried to manage their new lifestyle, and, again, the discipline of numbers came to their aid. The earliest problems reported by the Sumerians and Egyptians were written in prose. A simple example might be: if three loaves of bread are needed for one worker, then how many will be needed for a hundred and forty? Or, what is the total area if the parts are three acres, seven acres, and nine acres? This approach to stating problems is known as *rhetorical* algebra. The ancients used it because they did not have our modern symbolism to represent various quantities and operations. The rhetorical approach was slow and difficult, leaving room for different individuals to write the same problem in different ways. Hence, there was little standardization.

Modern symbolic algebra replaces the words with mathematical symbols, and in the process the problems are simplified and easily categorized, and solutions are more forthcoming. We can also use symbolic representation to characterize the different kinds of problems that have historically triggered a search for additional numbers. The symbols needed are exceedingly simple. We will use the four operational symbols: $+$, $-$, $\cdot$ (sometimes an x), and / for addition, subtraction, multiplication, and division. We will use letters from the end of the alphabet to represent numbers that are the solutions. We will use the equation as the format for our problems, with the number value on the left of the equation balanced by an equal value on the right. We will divide the equation between right and left with the equal sign ($=$). Later we will add a few simplifying conventions. For now, these few symbols, in conjunction with numbers and letters of the alphabet representing unknowns, is all we need.

To see how the rhetorical approach to problem solving translates into the modern symbolic approach we use the following example, which might have been found in ancient Egypt: If there are seven workers and each receives four measures of weak beer, and their foreman receives a double allotment, then how much weak beer is required?

First, we reorder the words into the form of an equation: the total weak beer needed is equal to four measures for each of seven workers plus twice four for the foreman. A major stage in going from rhetorical algebra to symbolic algebra is called syncopation, which was first used by the Greek mathematician Diophanus in the third century A.D. In syncopated algebra we abbreviate some of the words. Abbreviating key words we get: T.B. (or total beer) equals four measures for seven W. plus twice four for F. Now we will replace the word 'equals' with the equal sign: T.B. $=$ four measures for seven W. plus twice four for F.

We are getting closer to our symbolic notation as our problem statement gets shorter. We use the operational signs or $+$, $-$, $\cdot$, and $/$, and numerals for number words: T.B. $= 4$ measures $\cdot$ 7 W $+ 2 \cdot 4$ measures F. Now we will substitute a capital X for our T.B., since this is our unknown quantity. We can also delete the rest of the words since we know the answer is going to be in the number of measures of beer.

$$X = 4 \cdot 7 + 2 \cdot 4$$

Now, without the confusing clutter of the words, we can go about solving this problem by carrying out the two multiplications and the addition. The answer is, of course, thirty-six measures of weak beer. This rather simple problem seems easy to us, but it has not always been so straightforward. One can imagine the kinds of confusion and frequent arguments that must have occurred during the distribution of property, the paying of wages, or the collection of taxes in ancient civilizations when entire problems were stated in rhetorical form.

The basic algebraic problem, whether stated in rhetorical or symbolic form, is finding a number. To calculate this number we manipulate other numbers with the four operations of addition, subtraction, multiplication, and division. But, if we begin with numbers and then perform the necessary operations, do we always get another number for an answer? This difficult question plagued the ancients, because their number systems were too restricted to give them a useful result in all cases.

If we are certain that a specific operation used on numbers always results in another number of the same kind, then we say that the numbers are "closed" under that operation.

For example, the natural numbers are closed under addition. Why? Because whenever we add two natural numbers, we always get another natural number as an answer. However, the natural numbers are not closed under subtraction because it is possible to subtract two natural numbers and get something other than a natural number. For example, $3 - 7 = -4$, or three minus seven is a minus four. What is this strange -4? Whatever it is, it is not a natural number.

Closed numbers: A set of numbers is *closed* for an operation if every application of that operation on numbers from the set yields another number in the set.

We can use this simple idea of closure in conjunction with symbolic algebra to understand just where our ancestors were in their mathematical sophistication, and why they sometimes had trouble in understanding abstract relationships. If we consider different kinds of problems in the form of symbolic algebra, we can see how limited the Egyptians and Babylonians were. For example, the following simple equations would have been impossible for them to solve: $x + 7 = 4$. This equation yields -3 as a solution, a number which they did not recognize. Another would be: $x + 2 = 2$. This one, of course requires that x be zero, another number not included in their systems.

A simple notation which represents a number multiplied by itself is the exponent. For example, $7 \cdot 7$ is 7^2 and $7 \cdot 7 \cdot 7$ is 7^3. We can use exponents for unknowns too. For example, a problem facing early farmers might have been: If our land is square and covers 100 acres, then how many acres are on one side? The equation is: $X^2 = 100$ or $X = 10$ acres to a side. We chose the number 100 because the answer was so obvious. What if it had been 93 acres rather than 100? The equation would have been $X^2 = 93$, and the solution would have been $X = \sqrt{93}$. Both the Egyptians and Babylonians could approximate the answer, but neither realized that the exact solution required another kind of number. For this discovery, we must await the Greeks.

HOW FAR HAVE WE COME?

Previously we saw that the first numbers discovered were the natural numbers. They were primarily used to keep track of possessions by computing the cardinal number for a set of objects. As farming began and cities formed, our ancestors faced new problems. They had to count days to produce calendars, measure fields for planting, compute volumes of beer and barley, and somehow record all this. Writing was invented, number systems evolved, and fractions were born, while practical algebra and geometry got their start. Now people had two kinds of numbers: natural numbers and fractions. What more could they possibly need?

CHAPTER 5

Chinese and New World Numbers

Although modern mathematics and our modern concept of numbers are the products of Western and Middle Eastern societies, the Chinese and Native American cultures are still of interest. Both civilizations were isolated from the influences of other peoples, especially the Native Americans. Hence, we have the opportunity to review the Chinese and Native American concepts of numbers as independent developments. This allows us to consider whether our modern idea of numbers is heavily biased by their originators, primarily the Babylonians, Egyptians, and Greeks. In other words, do people in separate societies develop completely different number systems, or are the similarities greater than the differences?

ANCIENT CHINA

The ancient civilization of China was either contemporary with or somewhat younger than Mesopotamia and Egypt. While limited contact occurred between China and India, and possibly even with the West, it is uncertain which way the influence flowed. An analysis of China's early mathematics supports the view of independent development. Even accepting some mutual influence, it is reasonably certain that China's mathematics, for most of its history, have been isolated from outside cultures.

Some Chinese scholars have placed exceedingly ancient dates on China's first rulers, as old as 17,000 B.C. While modern (western) scholarship has revealed the richness and complexity of Chinese culture,[1] it has also shown that such ancient dates are exaggerated.[2] To add to the confusion surrounding China's history, scholars have difficulty placing various Chinese events in a fixed time frame before 1122 B.C., the reign of Wu Wang. The confusion on dates is caused in part by China's early internal strife which, at times, broke the continuity of the society. For example, in 213 B.C. the emperor, Shï Huang-ti, ordered the burning of books. We can only guess how successfully his order was carried out and what manuscripts have been lost forever.

A reasonable beginning for Chinese civilization is 2852–2738 B.C., which was the reign of Fuh-hi, reportedly the first emperor of China. His rule in China occurred after the unification of Upper and Lower Egypt, coinciding with the building of the pyramids and the Sumerian empire in Mesopotamia. During Fuh-hi's rule the Chinese were conducting extensive astronomical observations. Hence, their mathematics had already begun to evolve. Under the patronage of the Yellow Emperor, Huang-ti, who took the throne in 2704 B.C., it is reported that a text on astronomy was produced and a system based on sixty (sexagesimal) rather than ten (decimal) was established, although later Chinese mathematics was based on a decimal system. All during the third millennium mathematical activity continued.

The height of Chinese mathematics came during the thirteenth century A.D. when more than thirty mathematics schools were reported to be operating in the country. The overall level of sophistication achieved by the Chinese was superior to that of both the Babylonians and the Egyptians.

THE CHINESE NUMBER SYSTEM

Part of the difference between the Chinese number system and other systems is due to the structure of the Chinese language. Four hundred and twenty monosyllabic words comprise the Chinese lexicon, each of which can be spoken in four tones for a total of approximately 1700 sounds. To express the thousands of concepts, each word must have a number of meanings. In Chinese there are no tenses, genders, or articles.

On the other hand, the written language is based on over forty-five thousand individual pictograms or characters, each character sufficient to represent an entire idea.[3] Hence, spoken Chinese has few words (but many dialects) while the written language has many thousands of symbols. Therefore, the written language, based on pictograms rather than phonetic words, could be read by all the ancient literate people throughout the empire. This was a unifying force, not found in most other societies, which helped weld the Chinese people together, giving the government far-flung control.

The basic number system that evolved in China was decimal with unique words for the numbers one through nine. The word for ten was *shih* and began a new cycle. The word for one was *i* and the word for two was *erh*. The words eleven, twelve, and so on were formed by combining the word for ten with the smaller number words: eleven was *shih-i*, twelve was *shih-erh*. The decades, twenty through ninety, reversed the unit words and the *shih*: twenty was simply *erh-shih*. In addition were words for one hundred (*pai*), one thousand (*ch'ien*), and ten thousand (*wan*). Numbers into the millions were formed by combining the larger number words.

For example, the number 78,426 is formed by using the individual words for the units in combination with the numbers ten and up. The 7 (*ch'i*) is combined with ten thousand (*wan*) to get *ch'i-wan*. This is continued for the remaining digits until we have:

7	8	4	2	6
ch'i-wan	*pa-ch'ien*	*szu-pai*	*erh-shih*	*liu*

This spoken number system is very similar to our own English system. In English we say "seventy-eight thousand, four hundred twenty-six." In our system we make some combinations for numbers up to a hundred before attaching the larger place-holding numbers, for example, "hundred" or "thousand." The Chinese system is not a positional system since the different compound numbers can be moved about without losing their basic magnitudes. Notice that in English, too, the words could be moved around without losing the sense of the number, although stylistic rules would be violated. We could say six, four hundred, and seventy-eight thousand.

When the Chinese shifted from the spoken number word to the written numeral they retained their basic system. Each number in the spoken number word system is assigned a unique pictogram and the process of number combination is the same. In ancient China four different written systems evolved (three of which are shown in Figure 16): basic numerals, commercial numerals, official numerals, and stick numerals. Using the commercial numerals from Figure 16 we can compile the number 78,426 in the following way:

	Basic	Commercial	Stick	Name
0		○	○	ling
1	—	╎	—	i
2	ニ	╎╎	ニ	erh
3	≡	╎╎╎	≡	san
4	四	✕	≣	szu
5	五	𝟖	≣	wu
6	大	⊥	⊤	liu
7	七	⊥	π	ch'i
8	Λ	±	⊞	pa
9	九	𝄢	▥	chiu
10	✝	✝	—○	shi

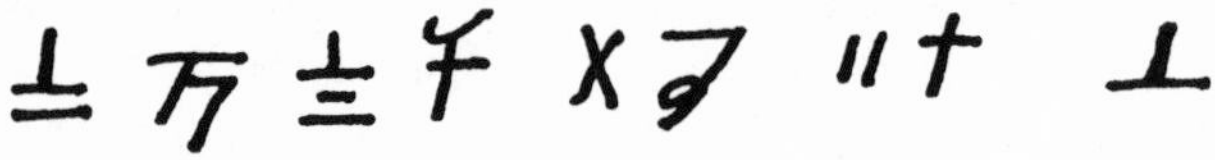

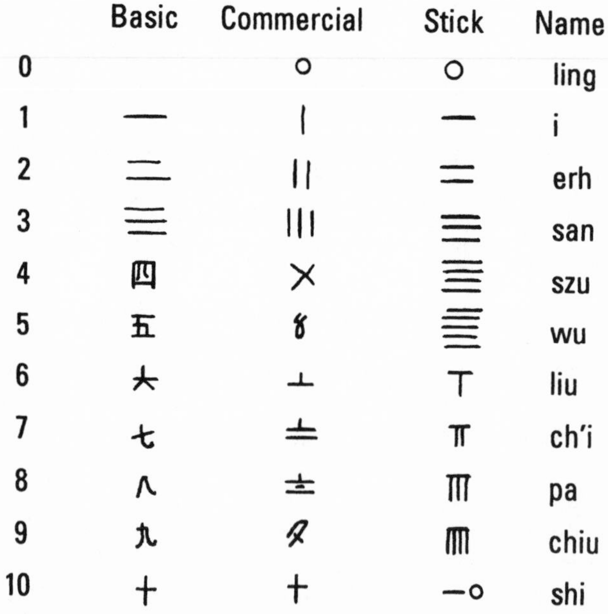

FIGURE 16. Three kinds of ancient Chinese numerals and their corresponding number words.

In Chinese, the number is written from top to bottom, rather than left to right. When the Chinese shift from the number word to the written numeral they still lack a position system such as our Hindu–Arabic system. A pair of symbols is used for each digit, one symbol to indicate the numeral and a second symbol to indicate the position value. Hence, it is possible to shift the positions of the pairs of the Chinese numeral symbols without loss of meaning (if one keeps track of the units digit). If such shifting of digits is attempted with our Hindu–Arabic system, the number changes.

The Chinese were not bothered by a lack of a zero since it was not needed. Just as we do not bother to say "four hundred, zero tens, and seven," the Chinese could ignore the zero in both the spoken and written language. It was not until the thirteenth century A.D. that a symbol for zero began to appear in their writing.[4] This was later than the appearance of the zero in both the European numerals and the New World Mayan numerals.

The first number words used by prehistoric human beings were less abstract and more concrete; they were used as modifying nouns to describe objects. In English we still have a brace of oxen and a pair of gloves. This primitive attribute is still evident in the Chinese language, which has approximately one hundred different number classes.

CHINESE MATHEMATICS

Four classic works survive from ancient China that help us understand Chinese mathematics prior to 1000 B.C. The first is the *Shu-king* or "Canon of History," which is attributed to the end of the third millennium and may have been written by Emperor Yau (ca. 2357–2258 B.C.). This work reports that two brother-astronomers, Ho and Hi, came under the emperor's wrath for failing to predict a solar eclipse. This ancient Chinese work involving astronomical calculations sophisticated enough to forecast an eclipse challenges those of early Greece some fifteen hundred years later. The second work is the *I-Ching* or "Book of Changes," which has been popularized recently in the West. It may have been written by Won-wang during the twelfth century B.C. The *I-Ching* is not really a book on mathematics, but is a book used for millennia by the Chinese to divine what course of action to take on important matters. Yet, it mentions the magic square, a square arrangement of nine numbers such that the three horizontal rows, the three vertical rows, and the two diagonals all add to the same number. The *I-Ching* method of casting oracles uses a combination of sixty-four hexagrams, and demonstrates that the Chinese were interested in the concepts of permutations and combinations. In all likelihood, such ideas were even older than the book, and may have been a part of Chinese culture for many centuries before 1200 B.C.

Mathematical manuscripts from ancient China that have survived are similar to the Babylonian and Egyptian works: They present the reader with a series of problems. The first truly mathematical text to survive was the *Chóu-peï*, which was probably written around 1100 B.C., and contains both calculations relating calendars and material on fractions. The work also includes a reference to dividing a line into the lengths of three, four, and five units, which most probably refers to the three-four-five right triangle. Therefore, the Chinese may have known about the Pythagorean theorem at this time.

The last of the four classic works, and one held in great esteem, is the *K'iu-ch'ang Suan-shu* or "Arithmetic in Nine Sections." It is thought to have been written by Ch'ang Ts'ang around 200 B.C., but based on much earlier works, some originating before 1000 B.C. It contains 246 problems divided into nine sections. The subjects include problems in taxation, surveying, percentages, computing the areas of triangles, circles, and trapezoids. It also has problems on simultaneous linear equations, the Pythagorean triangle, square and cube roots, and the use of the Rule of False Position, which we encountered in Egyptian mathematics. The Chinese not only solved equations involving square and cubes of unknowns, but simple equations with the unknown raised to the tenth power.

The Chinese solved indeterminate equations which they called *tai yen*. In a determinate equation there is one, or a small set, of correct answers. In an indeterminate equation there exist an infinite number of answers. For example, in modern notation we might have the equation $3X + 4Y = 17$. We can solve for X if we substitute a value for Y and, conversely, we can solve for Y if we substitute a value for X. This feature is what makes the equation "indeterminate," that is, the solutions are not specifically determined as the equation stands. Such equations have many real-world applications.

Probably the most significant contribution of the "Arithmetic in Nine Sections" to the theory of numbers is its mention of negative numbers. Here we have evidence that the Chinese at an early date, possibly before 1000 B.C., had stumbled onto an entirely new number, one not included in the natural numbers or the positive fractions. Yet, their acceptance of negative numbers was not complete, for even though they used them in calculations, they did not allow negative numbers to be solutions to equations.

As a whole, Chinese mathematics was superior to that of both the Babylonians and the Egyptians, but less sophisticated than the Greeks. While many of their efforts were directed to solving practical problems, they also dealt, to a limited degree, in proofs. They had a high regard for those who studied and performed mathematics. In fact, mathematicians were a necessary ingredient in the ancient Chinese courts because each new emperor ordered the Chinese calendar recalculated. This not only provided the court with the correct days for observing important rituals, but gave the common man a calendar for tracking the flooding of rivers, times for planting, and when to expect solar eclipses. Hence, it was up to the emperor, through the calculations of his astronomers and mathematicians, to demonstrate to the people that he was favored by Heaven and had a mandate from the gods to rule.

Chinese mathematicians had a great interest in computing ever better approximations for the value of π, the ratio of the circle's diameter to its circumference. The accuracy of this computation is a rough measure of a society's mathematical sophistication. Those early societies that did not embrace computation and mathematics generally used the value of 3 as close enough. For example, the ancient Hebrews used the value of 3 for π when constructing the temple of Solomon.[5] The Babylonian approximation was a little better. Sometimes they used 3 and at other times they used 3.125 which has an error of 0.52%. The Egyptian estimate of π was $\frac{256}{81}$ which is 3.1605 or an error of 0.60%.

Not only did the Chinese courts use mathematics, but it was necessary for ordinary merchants to be skilled in the basic arithmetic operations. Since the Chinese system was not positional, the written numerals could not be used in direct computation because the Chinese wrote down the place-value rank, making calculations with these ranks too cumbersome. Instead, the Chinese calculated mentally and then used a counting board to record the results. The counting board, which gave birth to the abacus, was a flat wooden surface with lines drawn to make a rectangle of squares. Sticks, about four inches long, were placed in different squares to represent units, the squares themselves representing larger numbers. Two kinds of sticks were used, red for positive numbers and black for negative numbers. It is reported that Chinese mathematicians developed great skill with their counting boards and could rapidly complete very complex operations.

The Chinese people also had a great interest in astrology, horoscopes, and predicting the future. Frequently, complex calculations were required to carry out the correct procedures. Hence, the general population had a real and immediate interest in mathematical mythology.

While the system of mathematics that evolved in China reached a considerable level of achievement, it had, unfortunately, little impact on the development of Western mathematics. When China finally opened her doors to allow modern technology to enter, her mathematicians had to catch up with their Western counterparts.

MATHEMATICS IN THE NEW WORLD

While some small influence may have passed between China and India, and even the Middle East, there is no evidence that the mathematics that developed in the New World was touched by any of the Old World societies. Many of the tribes of North and South America lived as hunter–gatherers, while others advanced to primitive farming right up to the time of contact with Europe. These tribes had no written language and, hence, no written mathematics. The kinds of spoken number systems used by these tribes ranged from pure 2-count and neo-2-count up to combinations of 5, 10, and 20-count systems.

The hunter–gatherers and the primitive farmers of North America generally counted on their fingers and kept tallies with sticks or pebbles. Some warriors kept track of the number of their battlefield victories with the notching and display of feathers. The Blackfoot Native Americans used a knotted string much like the Incan *quipu* to record tallies. The arithmetic was generally limited to addition through twenty and simple subtraction, while neither multiplication nor division were in evidence.[6] Such higher arithmetic activities may have been the responsibility of shamans or wise men. There is evidence that a number of large farming communities existed prior to the fifteenth century but which disappeared before the arrival of Europeans, especially in the Mississippi Valley with the Cahokia. What level of mathematics was achieved by these Native Americans, we cannot say.

Most North American Native Americans had no calendar. While they could specify the number of days since some past event or until an anticipated event, or even "moons" or years, they had no systematic procedure to track the cyclical events of nature, which would have allowed them a fixed record of past events or the ability to predict future events. Days did not accumulate into months, or weeks, nor did hours define the day. Evidence remains of pole rings used by both the Cahokia of Mississippi and the Native Americans of the Southwest for making astronomical alignments of the sun, and probably also the moon and major star groups. This shows that certain North American Native Americans were struggling to develop a calendar to forecast the seasons.

Why did the northern Native Americans develop little mathematical sophistication while at least two groups from Central and South America did? The most probable reason is that there was no need. Their way of life, as hunter–gatherers and primitive farmers, survived for millennia without those environmental pressures that demanded they develop more advanced arithmetic skills. While the farming villages of the Fertile Crescent were expanding, growing into cities, and supporting merchants and manufactured goods, Native Americans fulfilled their needs within the smaller tribal society. Yet, two New World societies did evolve: the Mayan and the Incan.

THE MAYA

The ancient Maya inhabited an area that includes modern Belize, Mexico's Yucatan, Guatemala, and western parts of Honduras and El

Salvador. As early as 9000 B.C. hunter–gatherers occupied the lowlands of Mexico's Yucatan and Belize. By approximately 2000 B.C. farming villages appeared in Belize.[7] The population grew, and around 400 B.C. large religious platforms began to appear. New theocracies evolved that were ruled by priests who supported a pantheistic religion based on a sun god. These were the Maya.

The Maya dug extensive canals to drain the swamps, and at the same time they raised the surrounding farmland with piles of soil. This allowed continuous cultivation which greatly increased their food production. The greater food production, in turn, caused another increase in population. It is estimated that the population of Tikal, one of the four religious centers, was between twenty thousand and eighty thousand at its height around A.D. 700, and that El Mirador, another Mayan city, may have supported as many as eighty thousand people.[8] Their society survived until A.D. 800, when they suddenly stopped erecting monuments and their cities began to depopulate. By A.D. 900 the Mayan society had fallen into decline and finally disintegrated about A.D. 1000 when its religious centers were invaded by the Toltecs from the north.

The Maya developed a sophisticated written hieroglyphic language that included both books and writings on monuments. The manuscripts contain characters that are simpler and more abstract than those on the monuments. To the great shame of the conquistadores, all but three books (now held in European cities) were destroyed.[9] One day in 1541 in the city of Mani on the Yucatan Peninsula, the Franciscan monk Diego de Landa burned the books of the Maya, forever destroying a priceless record of a great people. It is reported that de Landa later realized his crime against the Maya and spent his remaining life collecting accounts of the Mayan civilization from still living survivors of the society. Does the burning of these books remind us of other senseless acts of destruction? In 213 B.C. the Chinese emperor, Shï Huang-ti, ordered the burning of many of China's books; and in 48 B.C. Julius Caesar, while escaping from Alexandria with his army, caused much destruction to the great Alexandrian Library in Egypt, a library that included books from Rome, India, Greece, and Egypt. In A.D. 391 the Christian fanatic Archbishop Theophilus led a crowd that burned the rest of the Alexandrian Library, because the Christian emperor, Theodosius the Great, gave his permission for the destruction of all heathen temples. That so few men could cause the destruction of so much of the world's ancient heritage is beyond comprehension.

Most of our understanding of Mayan writing comes from 120 sites containing approximately eight hundred monument and building hieroglyphics, only five hundred of which have been translated. The Maya did not originate writing in Mesoamerica, for there is evidence that an earlier people, the Olmec, whose civilization lasted from 1200 B.C. until approximately A.D. 1, used a hieroglyphic script. In fact, all the advanced societies of Mexico and Central America used this hieroglyphic script along with the fifty-two-year double calendar, but it was the Maya who developed them to their most advanced states.

To date there are no pre-Spanish Mayan writings that directly deal with their mathematics. Therefore, our understanding of Mayan mathematics comes mostly from the numbers on monuments, and these monuments were used to record the dates of important events. We lack any record of their actual thoughts regarding mathematics, and our knowledge of their mathematics deals primarily with the Mayan calendar.

The oldest Mayan numbers appeared on monuments around A.D. 400. The Mayan numbering system was based on twenty, a vigesimal system, instead of a decimal system such as our own. Their system included both place values and a zero. Their numbers were written vertically, with the smaller value digits on the bottom and the larger on top. In the Mayan language there were unique number words for one through ten. The number words for eleven through nineteen are much like our English spoken number words—unit words combined with the word for ten. Remarkably, only three symbols were needed to write their numbers: a dot for the number one, a short bar for five, and an oval for zero. The written numbers between one and nineteen were formed from combinations of dots (ones) and bars (fives). The first five place values of our Hindu–Arabic place-value system are, of course

$$10,000 + 1,000 + 100 + 10 + 1$$

The first five Mayan place values were

$$144,000 + 7,200 + 360 + 20 + 1$$

These values are the result of the following multiplications:

$$20\cdot20\cdot20\cdot18 + 20\cdot20\cdot18 + 20\cdot18 + 20 + 1$$

Why use eighteen instead of all twenties? The logical system would be to use multiples of twenty at each step. However, their numbering system was used for recording dates. To build their calendar, they divided the year into eighteen months of twenty days each. This accounted for only 360 days, so they added another five days to fill the void; these five days were the *uayeb*—unlucky days. Hence, eighteen months times twenty days accounts for the eighteen in their place-value system. In their hieroglyphics, the value not only was indicated by its position but was accompanied by a grotesque pictograph (Figure 17).

An example of the Mayan written number 46,783 is shown in Figure 18. Written horizontally, with the place values indicated, we have

FIGURE 17. Mayan monument. The monument is on its side, yet the Mayan numerals of dots and bars are clearly visible along the upper edge. (Photograph from Brown Brothers, Sterling, PA.)

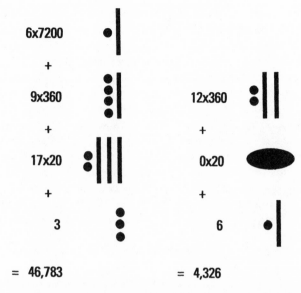

FIGURE 18. Mayan numerals for 46,783 (left) and 4,326 (right).

$$6 \cdot 7200 + 9 \cdot 360 + 17 \cdot 20 + 3 = 46,783$$

or

$$43,200 + 3,240 + 340 + 3 = 46,783$$

The second number in Figure 18 (4,326) shows the use of the Mayan zero in the second place. From Figure 18 we get

$$12 \cdot (18 \cdot 20) + 0 \cdot 20 + 6 = 4,326$$

This numbering system, based on twenty with an eighteen in the third position, seems needlessly complex. However, it was not intended for use by the common citizen, but was designed by the Mayan priests for their complex calendar. In all likelihood, the average person used the twenty spoken number words, and for those Maya who understood writing, possibly an elementary version of the dot and bar symbols for their every-day writing needs. The full system was reserved for the priest-rulers.

The motivation for such a complex number system was the Mayan religion, which worshiped a number of gods, including a sun god, a moon goddess, and probably a god or goddess for Venus, which they realized was both the evening and morning star. Tracking their heavenly gods required a good calendar, and to provide same, they actually created several calendars. The secular calendar, called *haab*, was the system of accounting for seasons with eighteen months of twenty days. This calendar was adjusted by adding five days each year, with additional adjustments (to account for the quarter-day each year) by observing Venus. An even more accurate year was calculated by the Mayan priests of Copan, who formulated the year at 365.2420 days, which is slightly closer to the true figure of 365.2422 days than our Gregorian calendar of 365.2425 days.

Another calendar was based on twenty named days in thirteen cycles for 260 days. This was the holy *tzolkin* calendar. The two calendars were combined so that a specific date required a numbered day within the tzolkin month plus a numbered day within the haab month. For both the secular and haab calendars to make a complete cycle took fifty-two years. This period was called a "calendar round." In addition to the haab, tzolkin, and calendar round, they used the "long count," a dating system that began with their date for creation. This long count was based on their number system and counted in periods of *baktun* (144,000 days), *katun* (7,200 days), *tun* (360 days), *uinal* (20 days), and finally *kin* (1 day). The Maya believed they were in the fifth cycle of the long count, which had a beginning date in our calendar of 3133 B.C. To suppose that the Maya began using their calendar at this remote time would be a fallacy. They probably assigned the date of 3133 B.C. based on legends involving their gods. Our date of 5 January A.D. 501 would be a total of $3,133 + 501 = 3,634$ years from creation. This is equal to 1,326,410 days (ignoring the extra days from leap years). To this we add the five days of January. The Mayan long count for this date is nine baktun, four katun, four tun, eight uinal, and fifteen kin.

In addition to the long count, which lasted over five thousand years or thirteen baktun, the Maya recognized a longer period of time composed of multiples of baktun called *alautun*. One such period lasted some sixty-three million years.

In summary, the Maya developed an extraordinary number system. Our Hindu–Arabic system may have evolved as early as A.D. 500, but more likely originated in India around A.D. 800.[10] Hence, the Maya had a positional system including zero four centuries before our modern system was developed, and a full millennium before the Hindu–Arabic system was adopted in Europe. In fact, the appearance of the Mayan zero probably took no more than a few hundred years, while in the Old World its appearance required several millennia.[11] Their calendar was more sophisticated than the European calendar of the same time period, and their science and astronomy may well have surpassed their contemporaries in Europe. They knew that Venus required 584 days to circle the sun and could track Venus's movements with an error of less than two hours in five hundred years.[12] One of their surviving manuscripts (the *Codex Dresdensis*) contains astronomical tables of great accuracy for both Venus and the moon. We can only wonder what treasures of Mayan mathematics were lost forever by de Landa's tragic burning of their books.

The writing and the calendar system of the Maya were adopted by later civilizations including the Toltecs, Mixtecs, and Aztecs. However, by the time of the Aztecs, the longer calendar periods had been abandoned for the fifty-two-year cycle, and the writing had deteriorated. The Aztecs no longer used the bar to represent five and had abandoned the zero.

THE INCAS

Around A.D. 1410 the Incas, who lived in the Cuzco valley of Peru, began to gain control over their neighbors. They continued to expand their boundaries until the arrival of Francisco Pizarro in 1532. The Spanish saw an empire stretching from northwestern Ecuador to central Chile, including parts of Peru, Bolivia, and Argentina, a distance from north to south of four thousand kilometers. The Inca population has been estimated at between six and twelve million inhabitants,[13] and accounts for the single largest New World empire. To compensate for any disruption of the food supply due to war or crop failure, the Incas established great storehouses throughout the kingdom and filled them with freeze-dried potatoes, called *chuño*. A system of highways, totaling some thirty thousand km, called the "highway of the sun" connected the capital at Cuzco to far-flung parts of the empire. In order to manage this huge empire, the Incas established a system of relay runners, called *chasquis*, to carry information from the provinces to the capital in Cuzco. To aid the relay runners, a system of stations were positioned along the roads approximately every 4.5 miles.

The Inca had no written language, and messages and official records had to be memorized. To aid in this memorization they developed a sophisticated system of knots tied on strings, called *quipus*, for recording numbers and the identity of items counted. Many other societies used such devices to record numbers at some time during their histories, including those of nineteenth-century Africa, ancient Greece, the Chinese, and the Hawaiians. The Incas constructed sophisticated and complex quipus (see Figure 10 in Chapter 4). Even though we have many quipus and can easily decipher how the numbers were recorded, the fine details of how they were used is lost to us since they had no written language.

The quipu was a number recording tool used by officials called *Camayoc* (or *Quipucamayoc*), who were responsible at many hierarchial levels within the government for recording almost anything that was of interest to the rulers. The Camayoc were trained in special schools open only to the elite of Incan society. The students were required not only to learn how to use the quipu but also to become familiar with the oral history associated with past quipu records. A quipu was made of a collection of strands of string, each string about 40 cm long. Some quipus contained over a hundred different strands, with each strand recording one number. Frequently each strand was colored to indicate which item was being counted. Sometimes one strand interlocked with others to show a total count.

The number system used by the Incas was based on ten. Each digit was represented by a combination of knots and was evenly spaced from the neighboring digits. A zero was indicated by a blank space on the string. Three kinds of knots were used: a single knot, a double knot, and a slip knot with from two to nine loops. The knots were tied in descending order on the string. Hence, the number 475 would have four knots on top, followed by a group of seven knots, and then either five single knots or a slip knot with five loops.

The Incas did not live in cities, as did the Maya and Aztecs, but occupied small villages seldom exceeding a thousand people. The capital and other provincial centers were temples where the bureaucracy lived. The decimal number system in conjunction with a 5-count was used to subdivide the population into groups of ten, fifty, one hundred, one thousand, five thousand, and ten thousand families. The farmland was divided into three sections, the produce of one section going to the temples, that from another going to the central leader (*Sapa Inca*), and the produce from the last section available to the farmer. Each citizen was required to devote a portion of his or her labor (women and children were included) to the government. All of this had to be recorded and accounted for by the Camayocs on their quipus, and the information, then transmitted to the center of power in Cuzco—a tremendous job.

Hence, the Incas, like the Maya, used a position-value system with a zero. However, where the Maya used their numbering system for developing their calendar to keep track of their gods, the Incas used theirs to control and manage the everyday activities of their immense population. The quipus were not used like an abacus to perform actual calculations. The Incas probably used fingers, pebbles, or other aids for doing calculations, then recorded the results on the quipus. It is also interesting to note the parallel between the Sumerians—the first to invent a token system to help manage commerce, croplands, and taxation, which then evolved into writing—and the Incan quipu system, which also helped manage people and possessions. Why did not the quipu system evolve into true writing? We must remember that it took the token system of western Asia approximately five and a half millennia (from 8500 to 3100 B.C.) to achieve this transition. The Incan Empire did not last even two hundred years! The achievements of the Incas, considering the short span of their empire, were truly stunning by any standard.

INDEPENDENT DISCOVERY

The similarities between the Chinese and New World number systems, and those of the Sumerians and Egyptians are striking. All the systems incorporated some combination of 5, 10, and 20 counting into their number scheme. Even the Sumerian system, based on sixty, contained a subdivision based on ten. Most of the systems contained position value. While the Fertile Crescent empires and China were slow in adopting a zero, all finally did so. Therefore, reappearing themes that occur in human number systems include a base that is related to counting on our fingers (five = one hand, ten = both hands, twenty = fingers and toes), a position-value system, and a zero.

We can now answer the question we posed at the beginning of this chapter: Are number systems that develop independently basically similar? The answer is yes. Human beings take the same general road to solving problems that require numeration and calculation.

The mathematics of the Chinese and Native Americans demonstrates that sophisticated number systems, both spoken and written, did not evolve as unique entities in the Fertile Crescent. In fact, if the Sumerians and Egyptians had never existed, we certainly would still have number systems and writing. What we see is the development of number systems in response to the problems encountered by farmers as their villages grew into towns and cities with the accompanying growth of manufacturing and trade. Wherever large groups of farmers collected, to be managed by a central power, new problems appeared that required advanced number systems to solve. Land had to be allocated, taxes computed and collected, armies raised and supported, goods shipped, and contracts fulfilled. Such problems created the accountant-scribes and accountant-priests, who in turn transformed the simplistic counting system of the primitive farmer into a tool to manage empires.

CHAPTER 6

Problems in Paradise

EARLY GREEK CIVILIZATION

Thus far we have reviewed the progression of the science of numbers in the Old World from prehistoric times to approximately 1000 B.C. We have seen the concept of number expand from the natural numbers to include the unit fractions (with form 1/n) of the Egyptians and the sexagesimal fractions (with form n/60) of the Babylonians, plus the negative numbers of the Chinese. It is now time to look at the contributions of the ancient Greeks. It has been claimed that the Greeks defined science, philosophy, and mathematics, and created standards for these disciplines that were maintained for two thousand years. Others, however, claim that we give too much credit to the Greeks, ignoring other societies' contributions.

Actually, both viewpoints are correct. The Greeks did redefine the practical calculating and mensuration of earlier civilizations into the sciences and mathematics we know today. The Greek achievements from approximately 600 B.C. to A.D. 300 overshadow all the gains in the intellectual arts for the next fifteen hundred years. But we must not be trapped into believing they worked in isolation without the benefit of the more ancient societies. They most certainly did borrow from both the Egyptians and the Babylonians. Yet, what they achieved was, in many respects, entirely new and characteristically Greek.

Greek civilization began approximately 2000 B.C. when a group of Greek-speaking, yet nonliterate, Indo-Europeans left the northern

Balkans and migrated into Greece and the Eastern Mediterranean.[1] There they began to assimilate the older culture which had been established on the island of Crete. For the next fourteen hundred years they spread to western Asia Minor (modern Turkey), the islands of the Aegean Sea and throughout Greece itself. The land they settled helped mold their varying forms of government. The Sumerians and later the Babylonians occupied a large flat plain: an easy area for rulers to move armies quickly, and an area where defenses were difficult to erect. This made it possible for powerful Sumerian and Babylonian princes to expand their rule and increase their territories. The Egyptians occupied a long valley, also an area that lent itself to collective rule.

But the Greeks settled in individual city-states that were usually surrounded by mountains and bordered by the sea. These cities offered ideal sites for defensive fortifications. Hence, it was difficult for any one Greek king to conquer his neighbors. Where the Babylonians and Egyptians had single rulers, the Greeks had alliances and confederacies of city-states. This meant the Greeks were not saddled with a single world view or a single cosmology imposed by priests through millennia of tradition. Each city-state could forge its own way. With the eastern Mediterranean crawling with these aggressive, curious Greeks, the world was ripe for an intellectual explosion.

Unified by their common background and language, the Greeks began the Panhellenic games in 776 B.C. at Olympia on the large southern Greek peninsula, the Peloponnese. From 775 to around 750 B.C. they began trading with the Phoenicians, who occupied a maritime state on the extreme eastern end of the Mediterranean, which was centered in the principal cities of Tyre and Sidon in modern-day Lebanon. The Greeks borrowed the Phoenician alphabet, which consisted only of consonants, and added vowels. With this new writing ability, the Greeks began recording their heritage of epic poetry. By 600 B.C., after a 150-year surge in migration, the Greeks occupied six hundred to seven hundred relatively independent cities stretching from the western shores of the Black Sea to southern Italy, Libya, and the Spanish coast. Around this time they began using coins in their commerce.

The Greeks now entered their classical period, which began in 600 B.C. and lasted until 300 B.C. The second period, the Hellenistic or

Alexandrian period, lasted from 300 B.C. until A.D. 600. However, our concerns are really with the classical period, for it was during this time that the Greeks defined their concept of number.

When the Greeks first began to use the alphabet during the eighth century B.C., they possessed no efficient medium to record their efforts. Parchment had not been invented, while clay and wax tablets were bulky and difficult to store. Fortunately, papyrus was introduced into the Greek world around 650 B.C. Now they had something on which to record their deepest thoughts, and they took full advantage of it. The earliest works of a scientific nature to survive came from the time of Plato in the fourth century B.C. The first substantial mathematics book is *Elements* by Euclid, who wrote around 300 B.C. In fact, no meaningful Greek mathematical manuscripts exist today. What we have are copies of copies of copies. The oldest of these copies are Byzantine manuscripts dating from A.D. 200 to 1200.

THE GREEK NUMBER SYSTEM

The Greeks used two kinds of number systems; the first is called the Herodianic or Attic system and was used from the earliest writing until it was phased out between 100 B.C. and A.D. 50. The second system, used predominately after 100 B.C., was the Ionic or Alexandrian system.

The Attic system was based on ten and used six primary symbols:

$$1 = |, 5 = \Gamma, 10 = \Delta, 100 = H, 1000 = X, 10,000 = M$$

While one was represented by a simple stroke, the other five symbols were all letters of the Greek alphabet. Various numbers were formed by combinations of these six symbols, much like the earlier Egyptian system and the later Roman system. The Greeks did not use position values as the Babylonians did or modern societies do. The symbol for five, Γ, was used in combination with other symbols as a multiplier. For example we have Γ^H which stands for 5·100 or 500. To get 5000 we simply combine Γ with X to get Γ^X. Other examples of numbers in the Attic system are

$$47 = \Delta\Delta\Delta\Delta\Gamma\|$$
$$374 = HHH\Gamma\Delta\Delta\|\|\|\|$$
$$23{,}621 = MMXXX\Gamma H\Delta\Delta\|$$

The symbols were usually written in descending order, but not always. You can see at once that order is not necessary in this nonpositional system. The Greeks did not use a zero. In general, the Attic system was not used for fractions.

The second system, or Ionic system, assigned letters of the alphabet to values as shown in Table 4. The Greek letters for the thousand series are identical to those for the units, except with an accent mark. Notice that Table 4 contains twenty-seven unique symbols, yet there were only twenty-four letters in the ancient Greek alphabet. Three extra symbols had to be added to round out the full twenty-seven: the digamma, F, for six; the koppa, φ, for ninety; and the sampi, λ, for nine hundred. The above letters are all capitals; the lowercase letters were not in use during the classic Greek period. With the above symbols, four or less were needed to represent every number from 1 to 9999. This represents a substantial improvement over the older Attic system. For larger numbers, several different methods were employed, including the myriad, M, from the Attic system for ten thousand. The myriad was then used in combination with the letters from Table 4 to generate even larger numbers.

The Ionic system could represent larger numbers with fewer letters

TABLE 4.   Greek Ionic Numerals

Units	Tens	Hundreds	Thousands
1 = A	10 = I	100 = P	1000 = ,A
2 = B	20 = K	200 = Σ	2000 = ,B
3 = Γ	30 = Λ	300 = T	3000 = ,Γ
4 = Δ	40 = M	400 = Y	4000 = ,Δ
5 = E	50 = N	500 = Φ	5000 = ,E
6 = F	60 = Ξ	600 = X	6000 = ,F
7 = Z	70 = O	700 = Ψ	7000 = ,Z
8 = H	80 = Π	800 = Ω	8000 = ,H
9 = Θ	90 = φ	900 = λ	9000 = ,Θ

than the older Attic system and was popular for use on coins. However, it was frequently desirable to distinguish the Greek numerals from Greek words since both were constructed from letters. This was accomplished by placing a bar over the numeral or sometimes by bracketing the letters with short rows of dots.

Fractions in the Ionic system were either unit fractions, like those used by the Egyptians, or proper fractions. Proper fractions were expressed in one of several ways: One method was an ordinary numeral for the numerator immediately followed by an accented numeral for the denominator, for example ΓΗ' for three-eighths. A second method was to write the numerator followed by the accented letter for the denominator twice. Hence, three-eighths would be ΓΗ'Η'. A third method was to show the numerator in normal position on the line with a small version of the denominator above it.

While both the Attic and Ionic systems were easy to read, neither system lent itself to computations. Since so many different symbols were in use, and conventions for identifying numbers or fractions varied, Greek numerals were confusing compared to our own Hindu–Arabic system. To see why, we need only consider a simple example. When we multiply ninety by three in our system we break it down into two simple steps. The first step is to multiply the zero of the ninety by three to get zero again. This we simply write down. The second step is to multiply three by nine to get twenty-seven. Now we write the twenty-seven to the left of our zero to get 270. What happens when we do the same easy problem in Ionic? We cannot multiply anything by zero because there is no zero in Ionic. We have to multiply three by ninety directly or Γ times ϙ equals ΣΟ. If we try to break the problem down by multiplying the Γ (3) by Θ (9) to yield ΚΖ (27), then we cannot get from the ΚΖ (27) to ΣΟ (270).

In the Hindu–Arabic system we can break the problem into easy steps, which we repeat. This is possible because ours is a place-value system. In the Ionic (and earlier Attic) system, we must memorize a much larger multiplication table. In our system the multiplication table is only nine-by-nine, which gives only eighty-one elements. In fact, it's really easier, since one multiplied by a second number always equals that second number. In practice we only have to memorize an eight-by-eight table or sixty-four elements. The number of different symbols in the Ionic system, coupled with the fact that it is a nonpositional system, requires a larger multiplication table. There is evidence that the Greeks actually memorized computational tables based on the sounds of the spoken words, rather than the written symbols. Hence, "seven times six equals forty-two" is easier to memorize than the corresponding symbols.[2] Like the Egyptians and Babylonians before them, the Greeks relied on the everyday use of tables for doing their calculations.

The Greeks most certainly used counting boards to actually carry out computations, yet there is little direct evidence of their method. The only surviving counting board from ancient Greece is a white marble tablet from the island of Salamis,[3] yet numerous literary references to them can be found. Counting boards, called *abákion* by the Greeks, could be made from various materials including wood, marble, or clay. Lines were drawn on the boards to define areas, and counters, called *pséphoi*, were moved about the boards to complete the operations of arithmetic. The actual details of day-to-day calculating are all but lost.

THE BIRTH OF PHILOSOPHY, SCIENCE, AND MATHEMATICS

Before the Greeks, philosophy, science, and mathematics were not defined fields of study. Ancient thinkers before 600 B.C. were priests, rulers, scribes, or merchants, but not mathematicians and scientists. The Greeks defined these words for us, and therefore helped shape our world of learning for all time.

On the west coast of Asia Minor was the city of Miletus, a wealthy trading center. Its merchant ships could easily reach the Nile of Egypt, while inland caravan routes connected the city to Babylon. At the same time, Miletus enjoyed trade with all the Greek city–states and the Phoenicians. It is here, at the crossroads of the East and West, that we meet the first great Greek philosopher and mathematician, Thales.

Thales lived from approximately 634 B.C. until 548 B.C. In his youth he was a merchant and is reported to have traveled to both Egypt and Babylon where he was probably exposed to the mathematics of both civilizations—civilizations that were by this time already several millennia old. In his later years he dedicated himself to the search for knowledge and founded the first Greek school, the Ionian School. This began a tradition of schools, not for tradesmen, children, slaves, or scribes, but for the adult male aristocrats of Greek society. A number of stories exist about Thales, but they are all secondhand and their authenticity must be questioned. Yet, they are of interest because they provide us with a general impression of how his compatriots and immediate successors viewed his contributions.

He was regarded by later Greeks as the first of the seven wise men of Greece. He has been given credit for originating Greek geometry, astronomy, and number theory. It was claimed that he greatly impressed his compatriots when he predicted a solar eclipse in 585 B.C., but modern scholars doubt that he actually did, since Greek astronomical knowledge was insufficiently developed at this time.[4] It was also reported that he once purchased all the olive presses in the vicinity of Miletus and neighboring Chios. When the picking season came, he rented them out at a high profit and thus made his fortune. He is the first man in history to have specific mathematical discoveries credited to him. Besides founding his school and teaching other Greek scholars, such as Anaximander and Anaximenes, Thales was reported to have taught the great Pythagoras, although it is more probable that Pythagoras attended his school after Thales had ceased to direct it.

Thales is given credit for a number of mathematical discoveries.[5] Four of these discoveries are illustrated in Figure 19. In (a) we see that the area of a circle is bisected into two equal halves by its diameter, that is, area A is equal to area B. In (b) it is shown that two triangles have the same size and shape (are congruent) if they have two angles and one side that are respectively equal. In (c) an angle is inscribed in the semicircle; Thales proved that all such inscribed angles must be right angles (90 degrees). The fourth discovery, shown in (d), is that the base angles, A and B, of an isosceles triangle are always congruent. Whether he actually made these discoveries is not important; his contribution is his method, not the specific results of application. His most meaningful contributions include logical deduction and abstraction.

When the Egyptians and Babylonians worked with geometry, they were measuring physical objects—distances on land, or the angles made

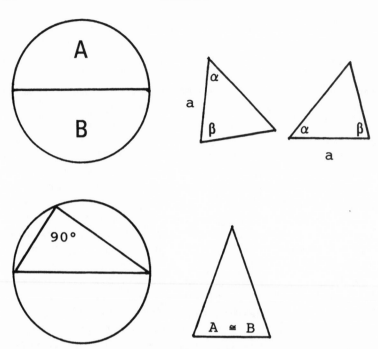

FIGURE 19. Four discoveries credited to Thales: (a) The diameter of a circle bisects the circle into two equal areas; (b) two triangles with one side and two angles congruent have the same shape; (c) a triangle inscribed in a semicircle is a right triangle; (d) an isosceles triangle has two congruent angles.

by intersecting ropes. Lines were physical markings drawn in the sand or ground. One of the great contributions of Thales was to speak of lines, circles, and other shapes in a purely abstract way. Lines were not something one could see in the sand, they were thought objects in our imaginations. This meant that abstract lines could be perfectly straight, and abstract circles could be perfectly round. He took the physical shapes and made them mental shapes. Then he did the next, most remarkable, thing. He began to make logical deductions that took him from one truth regarding his abstract shapes and led him to the discovery of additional truths. This method of demonstrating a geometric truth by logically deducing it from other truths so impressed other Greek thinkers that it began a general effort to discover all truth by logical deduction. This resulted in the establishment of a deductive scientific method in Greece that influenced Western thought for two millennia.

The Greeks demanded that their truths be absolutely certain; there was no room for error. The way they achieved this was by accepting premises that were self-evident, and then deducing conclusions that must, therefore, also be absolutely true. Hence, if their deductions differed from experience, then experience had to be abandoned in favor of the deductions. Modern science, on the other hand, relies on experience; it is experience that suggests our laws (these suggestions are called hypotheses). We then use logic to predict future experiences and verify the predicted experiences. Mathematical models may originally be pure creations, but to be useful in science, these models must relate to experience. In our science, if experience differs from the hypothesis or mathematical model, then we abandon that hypothesis and look for a better one.

While the deductive method works well for pure mathematics, used in isolation it is disastrous for science. Therefore, while the Greeks made many deductive discoveries regarding geometric relationships, their science was severely hampered by a defective method. This, in turn, resulted in some very strange conclusions about both reality and the basic nature of numbers. We might therefore say that Thales gave us a great gift in his deductive method for discovering geometric truths (in combination with his abstraction of geometric shapes), but later Greeks, so impressed with his method, mistakenly applied it to all kinds of truth, which retarded the development of science. Fortunately, there were some Greek thinkers who were willing to experiment and rely upon experience. Aristotle, during the fourth century B.C., made dissections to gather biological information, and Archimedes, working from the third century B.C., conducted elementary experiments in his search for the laws of physics. But it would be many centuries before these experimental approaches were allowed to blossom into modern science.

THE GREAT PYTHAGORAS

The stage was now set for one of the greatest mathematicians of all
time, Pythagoras (Figure 20). Various dates are given for his birth and
death, but a good approximation of his time would be 580 to 500 B.C. His
birthplace was the Island of Samos, not far from Miletus. Some claim he
studied directly under Thales, since their lives overlapped, while others

FIGURE 20. Pythagoras, ca. 580–500 B.C. (Photograph from Brown Brothers, Sterling, PA.)

say the differences in their ages makes this unlikely.[6] We do know that Pythagoras traveled extensively during his youth, visiting both Egypt and Babylon. Hence, he may or may not have been in Miletus while Thales was alive. In any case, he probably did study at the Ionian School where he would have learned Thales' deductive method.

In his later years, Pythagoras imitated the example of Thales and established a school at Croton in Magna Graecia (southern Italy). Pythagoras' group represented more than a school, it was really a religious cult of approximately three hundred rich and powerful young men from the surrounding peninsula. The religious rites included ceremonies to purify the soul, since Pythagoras believed the soul could survive after death and occupy other bodies, even those of animals. These rites also included mathematical elements. His disciples were divided into two groups, the inner group or *mathematikoi* (hence our word "mathematics"), who were privy to the teacher's most secret truths, and the outer circle called the *akousmatikoi* (or "those who hear"), who were schooled only in the cult's rules of conduct. Each mathematikoi had to first pass through the initiation stage as an akousmatikoi before being allowed into the inner circle. The members of the society were sworn to secrecy, which impeded the spread of their ideas. Pythagoras himself left no writings, so all our knowledge of his accomplishments are secondhand.

While the exact circumstances of Pythagoras' death are disputed, he did live to an advanced age. Afterward, because of local political upheavals, the school was disbanded. Some of the mathematikoi settled in Tarentum, also in Magna Graecia, where they continued their mathematical studies. Many of the akousmatikoi became traveling mystics.

Even with the abandonment of his school, Pythagoras' ideas continued to spread throughout the Greek world. His disciples established new schools and taught new students, and the Pythagoreans continued to have a major impact on Greek thought through the centuries. Plato, credited by many to be one of the world's greatest philosophers (a claim hotly debated by others), was heavily influenced by the Pythagoreans who lived during his time (Figure 21). Plato was so convinced of the importance of mathematics that he had the motto, "Let no one enter who is unversed in geometry" inscribed over the entrance to his Academy, a school that existed until A.D. 529.[7]

FIGURE 21. Plato, 427–347 B.C. (Photograph from Brown Brothers, Sterling, PA.)

Pythagoras' nonmathematical beliefs were a strange mixture of fact and fancy. His code of conduct seems to have been an odd assortment of silly rules: Do not eat beans, always stir the ashes after a fire. Yet, other beliefs were advanced, for example, he is reported to have maintained that the earth was a sphere.[8]

NUMBERS ACCORDING TO PYTHAGORAS

When we begin to look at the Pythagorean concept of numbers, we suddenly find ourselves in a web of convoluted thinking and flashes of brilliant inspiration. The basic calculating abilities of the Greeks were certainly in place by the time they began extensive trading with the Phoenicians, during the eighth century B.C. These skills were called *logistic* and were the tools of merchant and scribe. Logistic included the operations of addition, subtraction, multiplication, and division. It included manipulating both whole numbers and fractions. In addition, the Greeks were solving problems involving the distribution and accumulation of goods that today we would recognize as elementary linear algebra. All these skills were necessary to the merchants and scribes responsible for keeping trading goods moving and managing government affairs. Therefore, logistic was what we now call arithmetic. However, logistic was considered a skill or a trade, and not a truly intellectual pursuit. Hence, it was viewed in much the same light as fishing or carpentry.

On the other hand, the Greeks studied what they called *arithmetic*, the intellectual study of numbers. Arithmetic in the Greek sense was not our arithmetic, but what we now call number theory. They considered arithmetic to be subject matter worthy of true scholars.

This division of the study of numbers into arithmetic (logistic) and number theory (arithmetic) was unfortunate. It meant that arithmetic was ignored while only number theory was pursued. Yet, for previous millennia the human race's advancement in numbers had been driven by practical problems, and it was these problems that gave us the natural numbers and fractions. Now the Greeks decided to ignore the manipulation of these numbers (arithmetic) and concentrate elsewhere.

The Pythagoreans, in particular, were captured by number theory. It may have begun with a single discovery made by Pythagoras himself. What he discovered was the relation between the length of a taut string and the accompanying tone it produced when plucked. First, he saw if he decreased the length of a string to one-half of its original length, it produced a tone one octave higher. Tones one octave apart are "in harmony" and produce the most pleasing sensation to our ears. He then went on to discover those string lengths that produce the next harmonies; that is, a string reduced to three-quarters of its original length produces the fourth, two-thirds produces the fifth. Pythagoras asked what produced these harmonies between the various lengths of string. His answer had nothing to do with physics, that is, vibrations in the air and the acoustical design of our ears. Instead, he concluded that harmonies were purely the result of the ratios of certain numbers, for example, the ratios of numbers such as one to two, three to four, and two to three. From this, he made the inductive leap that all things depend upon numbers and the ratios between numbers.

Among the Pythagoreans we now see an explosion in the study of number relations as applied to virtually everything: Mathematics, science, astronomy, religion, politics—everything became dependent upon numbers. Yet, surprisingly, numbers for Pythagoras included only the natural numbers. Fractions did not count as true numbers for they were really ratios—the relationships between two whole numbers. "But wait a minute," you may be saying. "This is going backwards. We already have fractions from the Egyptians and Babylonians." Quite right, and old Pythagoras simply ignored what it took centuries to achieve. So why would anyone consider Pythagoras a great mathematician since he seems to have abandoned fractions to deal only with natural numbers? The story is just beginning.

We can now appreciate the connection between Pythagoras' claim—that everything depended on number—and the Greek marriage to deductive science. If all things depend on numbers, then, to learn all truths, all we have to do is deduce the truths about numbers. Hence, we begin with self-evident truths about numbers, deduce numerous conclusions, and these conclusions will contain truths about the physical world we live in. This process implies that deductive science does not need to study nature. All that experimental nonsense can be ignored. We only need to look at numbers and their interesting characteristics (number theory), and we can discover all truth.

Trying to bridge the gap between number theory and the real world caused the Pythagoreans to make some remarkably strange claims. In fact, they developed a kind of number mysticism that is still evident today in numerology. For example, odd numbers were masculine while even numbers were feminine. Perfect squares, such as four and nine, stood for justice; five stood for marriage—the combining of odd (male) and even (female). Six was the number of the soul and seven was understanding and health.[9] Heavenly bodies had to be located at distances that have definite ratios to each other and must share in harmonies. Hence, we have the harmonies of the heavenly spheres. Matter itself was literally made up of numbers. The very creation of the universe was caused by numbers. It began with the one (the Monad, the Limited) which was then separated by the principle of the Unlimited into the two (the Dyad).

To see how this number mysticism evolved it is instructive to review how the Pythagoreans manipulated "numbers." The Pythagoreans played with pebbles to form different shapes that revealed more truths about numbers. They probably derived this investigation from the ancient practice of calculating with pebbles on counting boards. The Greek word for pebble *pséphoi* is where we get our word "calculate." Some of the discoveries they made became cornerstones to modern number theory. For example, they discovered that certain collections of pebbles could be formed into triangles. Three pebbles made up the smallest such collection, and therefore, three became a triangular number (Figure 22). Other numbers that could be shaped into triangles were six, ten, and fifteen. Then they noticed that these triangular numbers (three, six, ten, and fifteen) were the consecutive sums of natural numbers:

$$1 + 2 = 3$$
$$1 + 2 + 3 = 6$$
$$1 + 2 + 3 + 4 = 10$$
$$1 + 2 + 3 + 4 + 5 = 15$$

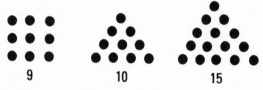

FIGURE 22. Pebble arrangements showing the four triangular numbers of 3, 6, 10, and 15, plus the two perfect squares, 4 and 9.

Therefore, one of the first discoveries of number theory was that the sum of consecutive numbers is a triangular number. If we look at the sum of consecutive odd numbers, we get square numbers.

$$1 + 3 = 4 = 2 \cdot 2$$
$$1 + 3 + 5 = 9 = 3 \cdot 3$$
$$1 + 3 + 5 + 7 = 16 = 4 \cdot 4$$

Prime numbers are natural numbers that can be divided evenly only by 1 and themselves. For example, two, three, five, and seven are all prime numbers. Composite numbers can be evenly divided by numbers other than one and themselves, one example being four which can be divided evenly by two. All natural numbers greater than one are either prime numbers or composite numbers. This distinction defines an entire field under number theory. The Pythagoreans knew the difference between prime and composite numbers, and Pythagoras was supposed to have searched for a test that would identify whether large numbers were either prime or composite. They discovered perfect numbers, or numbers that are the sum of their divisors. For example, 6 is the sum of 1 + 2 + 3.

The next perfect number is 28, or $1 + 2 + 4 + 7 + 14$. It is from these investigations that they founded the discipline of number theory.

The identification of pebbles or geometric dots with numbers made it possible for the Pythagoreans to identify physical shapes with specific numbers. Hence, the simplest two-dimensional figure, the triangle, became identified with three, and the simplest three-dimensional figure, the tetrahedron, was identified with four. In this manner, the Pythagoreans built up geometric solids which could then account for physical matter. In his *Metaphysics*, Aristotle remarked on the Pythagorean belief that numbers were, in fact, the very atoms of corporeal reality.

> They [the Pythagoreans] supposed the elements of numbers to be the elements of all things, and the whole heaven to be a musical scale and a number. . . . Evidently, then, these thinkers also consider that number is the principle both as matter for things and as forming both their modifications and their permanent states. . . .[10]

Later in the same book, Aristotle said,

> Again, the Pythagoreans, because they saw many attributes of numbers belonging to sensible bodies, supposed real things to be numbers—not separable numbers, however, but numbers of which real things consist. But why? Because the attributes of numbers are present in a musical scale and in the heavens and in many other things.[11]

Here we have the Pythagoreans advocating an atomic theory, where the atoms are made from points in space, which are themselves numbers. It is no wonder that they, and the many Greeks that they influenced, believed that the study of numbers would reveal all the secrets of the universe. While this belief in number–atoms may appear naive to us, we must remember what our modern hypothesis is: that matter is constructed of atoms, and that these atoms are almost entirely empty space! And when we try to pin down the other parts of the atom, namely electrons, protons, and neutrons, they dissolve into a plethora of eccentric objects that appear to violate our notions of space and time.

We may now ask, if numbers are collections of dots and occupy space, then how far apart are the individual dots? That is, if a tetrahedron exists in space (and is the number four), then how far apart are the points that represent its corners? The Pythagorean answer was that the relationships between lengths separating points had to be ratios of whole numbers. In other words, the various lengths of the sides of geometric numbers must compare to each other in whole number ratios. This is a critical point, and one which must have seemed obvious to the early Pythagoreans. It was only later, after Pythagoras' death, that this notion came back to haunt them.

At this juncture in our evolution of numbers we have a union of numbers with geometry. Numbers, for the Pythagoreans, were geometric, just as lines and circles were. In fact, the proofs given by the Pythagoreans for number theory discoveries were geometric proofs rather than algebraic proofs. Later we will see some examples of just how this was done.

PYTHAGOREAN CONTRIBUTIONS

One might be tempted to say that the silly Pythagoreans were just naive mystics who hurt the development of mathematics, rather than great thinkers of antiquity. But we must step back and take a broader view of their achievements. They made numerous mathematical discoveries, including a study of which geometric shapes have equal areas, the foundations to the theory of proportions; numerous theorems on triangles, circles, parallel lines, and spheres; and early discoveries in number theory.

Yet, their most important work was certainly on the right triangle and the theorem that we now call the Pythagorean theorem. This theorem is very simple; however, its importance to subsequent thinking is disproportionate to its simplicity. The theorem states that the sum of the squares of the lengths of the two adjacent legs of a right triangle is equal to the square of the length of the hypotenuse (see Figure 14 in Chapter 4).

One essential cornerstone of mathematics is the idea of a theorem. We may know a general rule for making calculations without having a proof that the rule always works. When it is possible to deduce a rule from other theorems or axioms we take to be true, then that rule becomes a theorem, that is, we know that the rule will always work. The idea of a theorem originated with the Greeks. Many pure mathematicians today consider their own work to be nothing more or less than the deduction of theorems from sets of axioms.

From previous material we know that Pythagorean numbers (three lengths that, fitted to the sides of a triangle, make a right triangle) were known to older societies. The Egyptian surveyors (rope stretchers) knew that a rope divided into the lengths of three, four, and five would form a right triangle (because $3^2 + 4^2 = 5^2$), but they probably did not know the general theorem. The Babylonians may have known the general theorem but probably did not have a proof of it. The Chinese, too, knew of the theorem before the Greeks yet offered no deductive proof. Therefore, while other civilizations knew the general rule of the Pythagorean theorem, they could not demonstrate that the rule always holds.

The genius of the Greeks is that they both knew the general rule and could also prove it, thereby raising the rule to the status of a theorem. Certainly the Pythagoreans had a proof of the Pythagorean theorem, and some think that Pythagoras, himself, gave such a proof. We will illustrate a proof provided by Euclid from his *Elements* without giving all the details; the details can be found in most mathematics history books.[12] Two good reasons exist for giving this proof: First, the theorem is important to the evolution of the Greek concept of number; and second, the proof demonstrates how the Greeks used geometric proofs. Figure 23 shows a right triangle with squares drawn on each side of the triangle, labeled as areas A, B, and C. Obviously we want to demonstrate that the sum of areas A and B is equal to C. This will give us $a^2 + b^2 = c^2$, which is what the theorem states. Euclid did this by dropping a line from the right angle down through C and parallel to the sides of C. He then proved that the area A was equal to the left rectangle in C while B was equal to the right rectangle in C. This gave the result that areas $A + B = $ area C.

A second, strictly visual proof is suggested by Figure 24. On the left side of Figure 24 we have one square inside a second square. The second, smaller square is equal in area to the square of the hypotenuse, c^2. The four triangles are right triangles with sides a, b, and c. We now take the four triangles from the left and rearrange them on the right. Notice that the two unshaded areas on the right are, in fact, a^2 and b^2. Hence, the unshaded area on the left is equal to the unshaded area on the right, which gives us $a^2 + b^2 = c^2$.

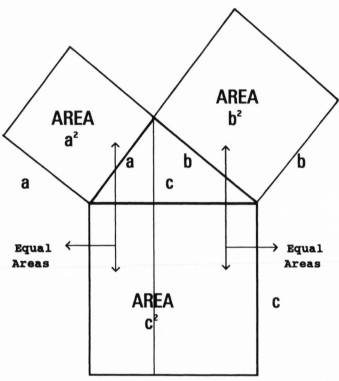

FIGURE 23. A proof of the Pythagorean theorem. The small rectangle in area c^2 is equal to area a^2 while the big rectangle is equal to area b^2. This demonstrates that $a^2 + b^2 = c^2$.

The Pythagorean theorem is one of the jewels of Euclidean geometry, which was taught as the standard geometry in the West for two thousand years. Yet, the importance of the Pythagoreans goes beyond their work in geometry and this particular theorem. What Pythagoras and his followers achieved was to make the study of science, and mathematics in particular, popular with the wealthy, propertied Greeks. Powerful Greek families wanted to send their young men to elite schools to learn what Thales and the Pythagoreans had discovered. This patronage

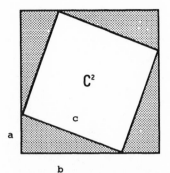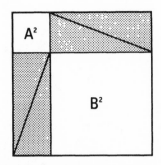

FIGURE 24. A visual proof of the Pythagorean theorem. By rearranging the four shaded triangles (left side) we change the area of c^2 into a^2 and b^2 (right side).

allowed scientific and philosophical thought to explode into a great golden age for the Greeks. Mathematical and philosophical ideas developed by the Pythagoreans influenced many later thinkers, especially Plato, the father of idealism.

Now we must consider another discovery about numbers due to the Pythagoreans. Yet, this discovery, rather than pushing forward the frontiers of number, set mathematics back for centuries.

THE INCOMMENSURABLE

The Pythagoreans believed that the ratios of the distances of geometric numbers (points) must be ratios of whole numbers. This is where the Pythagoreans got into trouble, and inadvertently discovered an entirely new class of numbers. Look at Figure 25. We will suppose that the sides a and b are each one unit long. How long is the hypotenuse, or side c? The Pythagoreans said that side c and the sides a and b must be in a ratio represented by two whole numbers. What could this ratio be? We can use the Pythagorean theorem to give us a symbolic representation for this number (that is, the length of side c).

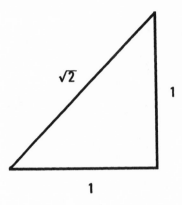

FIGURE 25. A right triangle with two legs having a length of 1. The hypotenuse has length $\sqrt{2}$, which cannot be written as any ratio of whole numbers, and is, therefore, incommensurable with the length of the legs.

$$c^2 = a^2 + b^2$$
$$c^2 = 1^2 + 1^2 = 2 \text{ hence,}$$
$$c = \sqrt{2}$$

But how long is this strange square root of two? We might be tempted to think this length, associated with a particular right triangle, rather obscure to worry about. However, it is a common value since it is also the length of a diagonal on a square with a side equal to one (Figure 26).

Certainly Pythagoras believed that two numbers could be found such that their ratio was equal to this value. Unfortunately, some Pythagorean, whose identity is now lost to us, proved after Pythagoras's death that no such natural numbers existed! That is, the length of c is *incommensurable* to the length of a or b. Whatever the ratio between c and a or c and b, it can never be exactly equal to a proper fraction. The Greeks called such lengths *alogos* for "inexpressible," or *arratos* for "having no ratio."[13] This idea destroyed a key feature of Pythagorean cosmology. How could everything in the universe depend on natural numbers and their ratios if such incommensurable lengths exist? These lengths obviously did not correspond to the way the Pythagoreans had defined numbers.

To say this discovery caused a scandal would be an understatement.

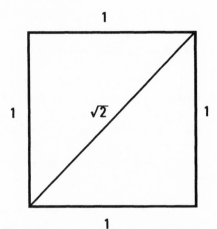

FIGURE 26. A square whose sides are each of length 1. The diagonal, with a length of $\sqrt{2}$, is incommensurable with the lengths of the sides.

Many different stories exist about what happened, and not all of them can be true. Some ancients gave credit for the discovery to one Hippasus of Metapontum who lived during the fifth century and was a member of the Pythagorean society.[14] One story says that Hippasus revealed the existence of incommensurable lengths to outsiders, thus violating the society's rule of secrecy. Some have said Pythagoras drowned Hippasus for his treason, but this is very doubtful, as the great teacher had most probably died by this time. Other stories say that the discoverer so angered the gods that they caused him to drown at sea.[15] Another version is that the Pythagoreans threw him overboard to drown him, or that they drowned him in a lake. Wherever the truth lies, these stories demonstrate how disastrous this discovery was to the Pythagorean idea of number—that is, the belief that numbers were physical and had extension in space.

The proof that the square root of two is incommensurable is simple, and because of the importance of this discovery we present it here, a proof that was preserved for us by Euclid. Let us suppose that there do exist two numbers that satisfy our ratio. We will call these numbers p and q. Therefore, p/q is the ratio and the value of $\sqrt{2}$. Let's also say that we have factored both numbers, p and q, into their primes and removed any common terms. It follows that both numbers cannot be even, since then they would both contain the common term of two. Therefore, we know from the start that if one of these numbers is even, the other must be odd.

We have assumed that $p/q = \sqrt{2}$. If we square both sides of the equation we get $p^2/q^2 = 2$. Multiplying both sides of this equation by q^2 gives us $p^2 = 2q^2$. Now, since p^2 is equal to 2 times another number, then p^2 must be an even number. If so, then p must be even for, if 2 is a factor of p^2, then 2 must be a factor of p. If p is an even number, then q must be odd, because both cannot be even.

If p is even we can represent it as $2r$. If we substitute $2r$ for p, we get $(2r)^2 = 2q^2$. This gives $4r^2 = 2q^2$, and, factoring out 2 from each side, $2r^2 = q^2$. Hence, q^2 is an even number, that is, it is equal to 2 times another number. But this means that q must be even also. But we have already shown that q must be odd. Therefore, the assumption that p and q exist results in the contradiction that both cannot be even, but that both are even. Therefore, p and q do not exist. Hence, there are no two natural numbers whose ratio is the square root of 2.

This method of assuming the opposite of what is to be proven and then showing that it leads to a logical contradiction was a favorite method of the Greeks, although they were not the first to use it. They called this method *reductio ad absurdum*.

It was not bad enough that this unknown Pythagorean discovered one incommensurable length. By about 400 B.C. a general proof was developed by Theaetetus (420?–369 B.C.), who studied under Plato, that the square root of any natural number, which is not itself a perfect square such as four or nine, is incommensurable relative to one. Hence, all of the magnitudes represented by $\sqrt{3}$, $\sqrt{5}$, $\sqrt{6}$, $\sqrt{7}$, and so on, are incommensurable with one. Pythagoras, because of his discovery of the relationship between whole number ratios of string lengths to musical notes and harmony, believed that all creation was based on whole number ratios. This idea became a cornerstone to the entire Pythagorean cosmology. The proof of the existence of geometric lengths that were not in whole number ratios dealt a death blow to the Pythagorean cosmology and physics. Yet, it also put the Greeks on the verge of discovering a new number, numbers we now call *irrational* numbers. But they were so horrified by the implications of this discovery that they took a step backward. Where the Pythagoreans claimed the unity of numbers and geometric extension, the later Greeks denied it. Hence, one could not associate numbers with lengths. This caused an unfortunate split between the disciplines of geometry, which dealt with points and lines (extension), and algebra, which dealt with numbers. This split lasted for two thousand years until René Descartes finally put them together again in analytical geometry. We can see this split reflected in the fields of mathematics as defined by the Pythagorean Archytas: They were music, geometry, astronomy, and number theory. Where was algebra? It was effectively ignored. Yet, this quadrivium of subjects defined by Archytas was adopted by Aristotle and remained the curriculum of mathematical subjects until the Renaissance.

THE ALEXANDRIAN PERIOD

We cannot assume that the bulk of Greek mathematics was developed in the classic period from 600 to 300 B.C. Although this period defined how the Greeks would think of numbers, it did not produce the great body of Greek geometry. This occurred during the second period, the Alexandrian period.

After the death of Alexander the Great in 323 B.C., his generals divided up his kingdom. Ptolemy ruled in Egypt. Earlier, Alexander himself had founded the new city of Alexandria on the Nile delta. In Alexandria Ptolemy began a library and school and invited one of the world's greatest mathematicians, Euclid, to teach there. A number of outstanding mathematicians followed Euclid to Alexandria, and their work added many theorems to geometry. Much of this work is included in the great geometry book by Euclid, *Elements*. Even in the much-neglected field of algebra, contributions were made at a later time when Diophantus published his book, *Arithmetika*, during the third century A.D. Euclid, Archimedes, Apollonius, and Diophantus made the major Greek contributions to both geometry and algebra. Yet, these advances did not expand the theory of numbers. For this we must redirect our attention to the East and the Hindus.

How can we summarize the Greek contribution to numbers? Certainly, the work by Thales, Pythagoras, and the Pythagorean school does not represent the bulk of Greek deductive mathematics. Yet, the tone of the investigation was established during the classical period, and specifically, the door that had been opened to irrational numbers was slammed shut.

In the final analysis we must recognize the Greek contributions, including the abstraction of numbers and geometrical figures, the use of deductive proofs to establish theorems, and the great many discoveries in geometry. And, to be fair, we must point out their deficiencies. They thought deduction sufficient for all truth, including empirical truth. They abandoned fractions as true numbers, and they divorced algebra, or the symbolic manipulation of numbers, from geometry.

CHAPTER 7

The Negative Numbers

Here we will survey several topics spread over a period of four thousand years. As we have seen, the concept of number became stifled in Greece toward the end of the classical period, actually going backward with respect to fractions. To rescue numbers, we will seek out the development of the zero, negative numbers, and our modern Hindu–Arabic numbering system, all of which contributed to our modern theory of numbers.

THE NUMBER LINE

Before we begin our exploration we are going to digress to a topic that will help us understand all the various kinds of numbers we will encounter. Previously we mentioned how the Pythagoreans tried to tie numbers directly to geometrical forms and failed. Now we will make the same attempt. We are going to demonstrate numbers by assigning them locations on a line. "But," you might protest, "isn't that just what the Pythagoreans did, and look what happened to them." You would be right, but we are not going to confuse numbers with geometric points. We understand that they are different. However, by associating numbers with points, we can better illustrate different kinds of numbers and how they are related.

We will begin with zero and the natural numbers. Let's take an ordinary straight line, as in Figure 27a. We pretend the line extends infinitely in both directions, while we see only the section in the figure.

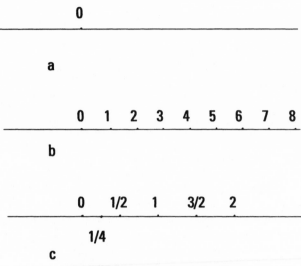

FIGURE 27. Building the number line.

Now we select one point and assign the number zero to that point, just as we have done in Figure 27a.

Before we go on, let's agree on exactly what we mean by lines and points. The physical picture of a line you see in Figure 27a is not the line under discussion. We are talking about an imaginary line in our thoughts. The line in Figure 27a is meant only to help us visualize what is meant. The physical line has width (else you could not see it), but the line in our imagination has no width. Also, the line in our imagination goes on forever (even though we do not "see" it doing this in our mind's eye) while the physical line is confined to the page. We can say the same for the dot in Figure 27a, which represents a point. Our imaginary points have neither width nor height. They represent only a location on the line. We use physical dots in Figure 27a only to communicate what is meant and help us to visualize the situation. Again, abstract lines have no width, and abstract points have no width and no height. They are simply locations.

Now we can proceed to build up our number line. Since we have zero located on the line, we are going to move to the right a little and find another point for the number one (Figure 27b). How far we move does not really make any difference in the way we construct our number line, but for the convenience of adding more numbers we have moved only about one-half an inch. Next we move the same distance to the right and select a point for the number two. We continue in this manner until we have selected points for the numbers from zero to eight. We will pretend that all the natural numbers have been assigned different points, each point one unit distance to the right of its predecessor.

We have assigned a point for every natural number and zero. But we know the Babylonians and Egyptians included fractions as numbers, so we assign points to fractions too. In Figure 27c we magnified the number line in Figure 27b so that we can see the zero, one, and two. Between zero and one we have positioned the fractions one-half and one-fourth. We did this by moving to the right of zero half the distance to one and assigning that point to one-half, and moving one quarter of the distance to locate the fraction one-fourth. This is how we will place all the fractions: We will simply move out the corresponding distance from zero and assign the appropriate point. For every natural number, there is a corresponding unit fraction. You can see at once that all the unit fractions of the Egyptians fall between the points one-half and zero on the number line. Of course, the Egyptians could represent other fractions as the sum of unit fractions, but, with the exception of two-thirds, they always used unit fractions. Possibly they used a special symbol for two-thirds because of its frequent occurrence.

While our number line extends for an infinite distance to the right to account for every natural number, the infinite number of unit fractions are located between zero and one-half. That is a lot of points crammed into a small space. But remember, these points we are assigning to our fractions do not have any width, so there is plenty of room for all of them.

We can locate all proper fractions on our number line. For example, to locate three-halves we simply divide the distance between zero and one into two equal parts and then move a distance equal to three of these parts to the right of zero. Three one-half lengths to the right places us exactly halfway between the one and the two in Figure 27c. Using this procedure

we can place all proper fractions on the line, and the entire line right of
zero becomes dense with fractions. In fact, the fractions will be distrib-
uted everywhere we look to the right of zero. That is, no line section, no
matter how small, will be free of fractions. We can imagine that we have a
magic magnifying glass which we hold over the line. When we use it to
magnify a segment of the line, we see this magnified line full of fractions.
Magnify it again and again, and we always see more and more fractions.
There simply does not exist any small part of the line that is free of
fractions. This idea of number density is a critical point that will be more
fully developed later and will prove useful in the future.

What about all those points left of zero? Can we assign numbers
here? Yes, for this is where we will place the negative numbers.

NUMBERS IN INDIA

Dates and events in early mathematics on the Indian subcontinent
are almost impossible to trace. Yet, we do know that an indigenous
civilization of great antiquity sprang up there and developed both writing
and the art of numbers. The oldest evidence is from the archaeological
site at Mohenjo Daro in the Indus Valley of modern Pakistan.[1] This
civilization goes back to the middle of the third millennium B.C., which
corresponds roughly with the construction of the Egyptian pyramids. The
inhabitants of Mohenjo Daro developed writing and a decimal number
system. Unfortunately little else is known about their mathematics.

Around 1500 B.C. the Indo-Europeans known as the Aryans migrated southeast, invaded India, and established a Hindu state based on
castes with the Brahmans at the top. The Hindu Brahmans wrote in
Sanskrit and kept all knowledge within their privileged upper class. This
secretive approach to learning stifled the recording and spread of mathematical ideas. It was not until the arrival of Buddha (560–483 B.C.—
during the life of Pythagoras!) that the monopoly of the Brahmans was
challenged by the new Buddhist religion, and literature began to flourish.
Unfortunately, no mathematical manuscripts survived that period. Our
first mathematical works from India appear in the fifth century A.D.
However, other Hindu literary works indicate that some mathematics was
produced during the previous millennium.

From 800 B.C. to A.D. 200 a number of religious Hindu manuscripts were written which are collectively called the *Sulvasūtras*. Rules in the Sulvasūtras prescribed how each family head was to construct his family altar.[2] The most common shapes were circular, square, and semicircular. However, each shape had to have the same specified size. This led to the evolution of a primitive geometry involving equal areas. For example, as early as the fourth or fifth century B.C. we find an approximation for $\sqrt{2}$ without any indication whether the author knew it was only an approximation. Also found in the Sulvasūtras are references to Pythagorean triplet numbers, such as three, four, and five. Unfortunately, not much more is known about this period.

After the Sulvasūtras period ended in A.D. 200 we have the *Siddhāntas* period. Included in this period is the *Sūrya Siddhānta*, India's first known work on astronomy. Much of the work from this point on was driven by problems in astronomy, with many of the important mathematicians appointed as court astronomers. The first known Hindu mathematician, *Āryabhata*, was born in A.D. 476. His writing shows the earliest evidence for a decimal place-value system. Āryabhata also used trigonometry in spherical measurements, and his estimate of π, 62,832/20,000 or 3.1416, is also interesting. The next contributor was the astronomer *Varāhamihira* (ca. 505) whose work included those computations necessary to find the position of a planet. He also taught that the earth was a sphere; however, this was many centuries after the Greeks had made the discovery.

During the seventh century A.D. we have the prominent mathematician, *Brahmagupta* (ca. 628) who contributed in the areas of number sequences or progressions, interest rates, mensuration of geometric areas and volumes, and algebra used for astronomical calculations.[3] However, it is his work on negative numbers and zero that is of particular interest, for he wrote the first comprehensive rules for computing with both. Brahmagupta's contribution on negative numbers is frequently ignored or glossed over in history books. While the possibility exists that earlier mathematicians postulated the existence of negative numbers and may even have worked out the rules for manipulating them, we must give credit to Brahmagupta. The early Greek mathematicians, including Pythagoras, damaged the concept of numbers by rejecting fractions and negative numbers; the Hindus, beginning with Brahmagupta, avoided such damage by accepting both. Unfortunately, it took the Europeans centuries to progress to the same acceptance of negative numbers.

Mahāvīra (ca. 850) also accepted negative numbers and zero and restated the rules governing their use. The last of the great Hindu mathematicians, *Bhāskara* (1114–1185), added a new twist when he treated negative quantities as debts or losses and the positive numbers as assets. This added a common application for the use of negative numbers that we still find useful today. In fact, it was probably the Hindu mathematicians' ability to see negative numbers as debts that encouraged their final acceptance, for when we can demonstrate that a mathematical concept has a useful application, we prepare ourselves for willing acceptance of that concept. Bhāskara even talked of negative numbers as roots or solutions to equations. Although the acceptance of negative numbers was not universal in India, it did increase over time.

In conjunction with the use of zero and negative numbers, the Hindu mathematics evolved from the primitive rhetorical algebra, which uses no symbols, to an almost completely symbolic algebra, achieving a level of symbolic abstraction beyond that achieved by Diophantus, the best algebraist of the Greeks.

The acceptance of the zero by the Hindus parallels their use of negative numbers. However, some authorities[4] have suggested that the Indians actually obtained the idea of zero from the Greeks. During the first half of the second century A.D., the Greek mathematician Ptolemy of Alexandria wrote a manuscript composed of thirteen books that contained the foundation to what would become, in modern times, trigonometry. In his work, Ptolemy used an "o" to represent a missing value. It may have been the first letter in the Greek word *ouden*, meaning "nothing." The suggestion has been made that the Indians adopted Ptolemy's symbol for the zero, leading to the question of outside influence on Hindu mathematics.

OUTSIDE INFLUENCE

Different scholars take various positions on how much the Indians were influenced by outsiders, especially by the Greeks, and how much was indigenous development. For example, from John McLeish, Emeritus Professor at the University of Victoria, British Columbia, we have, "There is no evidence that the Greeks had any influence on Indian mathematics."[5] On the other side, we have the viewpoint of Morris Kline, Emeritus Professor of Mathematics, Courant Institute of Mathematical Science, New York University:

> The second period of Hindu mathematics, the high period, may be roughly dated from about A.D. 200 to 1200. During the first part of this period, the civilization at Alexandria definitely influenced the Hindus. . . . The geometry of the Hindus was certainly Greek. . . .[6]

Certainly, the Hindus had to be aware of the Greeks, since Alexander the Great conquered the Persian Empire and marched all the way to the Indus Valley between 327 and 325 B.C. Therefore, there must have been some opportunities for Greek and Hindu scholars to exchange ideas. We also have the question of influence from both Babylonia and China. In all likelihood there was influence, yet this does not diminish the contributions of the Hindus. Regarding the zero, there is no direct evidence that they did get it from Ptolemy's school. It is also useful to remember that the Mayans, independent of any outside influence, developed a zero not unlike the Hindu zero. This suggests that representing zero as an "O" or a small circle may be a natural way for the human brain to symbolize "nothing."

THE HINDU–ARABIC NUMBER SYSTEM

The greatest contribution coming from India turns out not to be their acceptance of a zero or negative numbers, as meaningful as these contributions are, but rather their unique number system, which is now used worldwide. This number system contains four elements: a system based on ten; a set of unique symbols for the numbers one to nine; a place-value notation, and a zero. None of these is alone unique to India, for we see these elements existing separately, or in smaller combinations, in other civilizations. The Mayans had both a place-value system and a zero. The Greeks had a ten-base system and unique symbols for one to nine (in the

Ionic system), but no place value and no zero. Therefore, it was the combining of all these elements that gave the unique quality to the Hindu system.

As early as the third century B.C. a set of unique symbols for one through nine appeared on inscriptions.[7] These are the Brahmi numerals, which had not yet evolved into the symbols we use today (Figure 28) and did not include the zero. When the Brahmi numerals were first used, they were simply added, much like the Greek or Egyptian numerals, and did not use a place value. Around A.D. 600 the Brahmi numerals began to appear in a place-value system. At this time, or soon after, the zero was added to make the system complete. It is a slightly altered form of this system that was transcribed by the Arabs and found its way into Europe.

The major Indian contributions are in two areas: the acceptance of

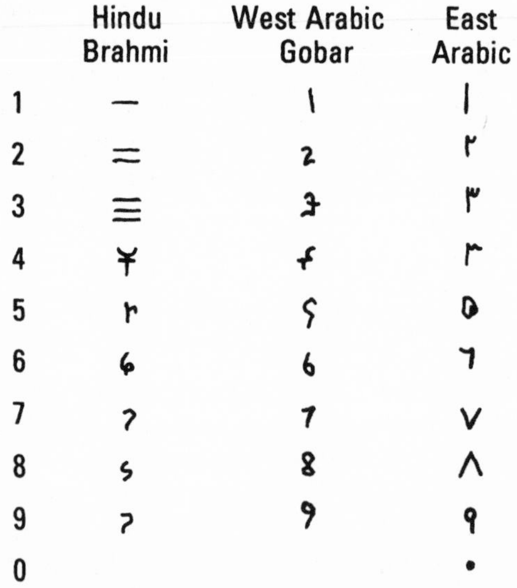

	Hindu Brahmi	West Arabic Gobar	East Arabic
1	—	١	١
2	=	٢	٢
3	≡	٣	٣
4	✶	✔	٣
5	٢	٢	٥
6	٤	٦	٦
7	٢	٢	٧
8	٢	٢	٨
9	٢	٩	٩
0			•

FIGURE 28. Precursors to the modern Hindu–Arabic numerals. On the left are the Brahmi numerals from India (third century B.C.), and in the middle and right are the two Arabic numerals (ninth century A.D.).

negative numbers and zero, and the combining of the four separate elements to make our modern number system. It was the second contribution that had the greater impact, although both contributions were only slowly accepted by Western mathematicians. The Indians also made a significant contribution in designating negative numbers as debts and thus identifying them with an everyday application.

By 1300 the Hindus ceased to produce any great mathematicians. It was not until the twentieth century that we again encounter Indian mathematical genius in the name of Srinivasa Ramanujan (1887–1920).

THE NUMBER SYSTEM GOES TO EUROPE

The modern number system is called Hindu–Arabic because the Hindus developed it and the Arabs preserved it and introduced it into Europe—by transcribing the Hindu manuscripts and preserving them for others.

During the seventh century, the Prophet Muhammad united the nomadic Arabs of the Arabian peninsula. In a single century the Arabs conquered an empire from India to North Africa and Spain. This great empire divided into a western kingdom with its capital in Cordova, Spain, and an eastern kingdom centered around Baghdad, which was founded in 766 by Caliph Al'Mansur.[8] The Caliph opened the gates of Baghdad to all scholars, regardless of ethnic or religious origin, and Baghdad became a great center for learning, and a repository of knowledge, especially that of the Greeks and Hindus.

Several Arabic mathematicians translated Indian manuscripts containing the Hindu numerals, including Mohammad ibn Musa al-Khowarizmi who wrote on Indian arithmetic in 830. The Arabs altered the Hindu numbers into two numeral sets, the *Gobar* (*Gubar*) or *West Arabic* numerals and the *East Arabic* numerals (Figure 28).

After the fall of the Greek and Roman empires, the light of learning flickered out in Europe. From the end of the fourth century to the Renaissance, there were no meaningful advances in mathematics. Greek Ionic numerals were used in Europe until they were replaced by the Roman numerals during the tenth century. The abacus and counting board were used by merchants for calculating. Gradually, however, cracks in this great edifice of ignorance began to appear. The first attempt to introduce the Hindu–Arabic numbers into Europe was made by none other than Pope Sylvester II during the tenth century. It was not successful.

There were good reasons to adopt the new numerals, but there were also strong pressures to keep the old Roman system. The main advantage of the Hindu–Arabic numerals was that they could be used to perform actual calculations. That is, instead of using an abacus or counting board, the scribe could write the numbers down and then manipulate the individual numerals, just as we do today when we stack numbers to add, subtract, or multiply them. In the older system, the scribe manipulated beads on the counting board or on the abacus, and, when finished, wrote down only the answer on paper or vellum. To simply record a number, the Roman numerals worked just fine.

The advantage of writing out the steps to a problem on paper rather than using an abacus is that someone else other than the author can see how the answer was obtained, even at a much later time. Hence, the new numerals allowed auditors to check calculations. This was definitely an advantage that helped control fraud and theft. However, a great disadvantage to writing the problem out was that it consumed paper, and paper was both expensive and rare. A second medium in Europe was vellum made from hides, which was also expensive. In other words, the medieval Europeans did not have tons of cheap paper for scribbling. By using a counting board or abacus, the scribe saved paper and vellum for only the most important part of the computation, writing the answer. The availability of paper can be appreciated when we realize that Europe's first paper mill did not appear until 1154 in Spain, while Baghdad had one as early as 794.

The next attempt to introduce Hindu–Arabic numerals into Europe was carried out by the one outstanding mathematician of the Dark Ages, Leonardo of Pisa (also known as Fibonacci, ca. 1180–1250).[9] In 1202 Leonardo wrote *Liber Abaci*, a book on computing which contained the new numerals. For centuries his book was used as a source book on calculation. A second push for the new numerals came from Johannes de Sacrobosco, who taught at the University of Paris. In 1240, he wrote a text on using the Hindu–Arabic numerals in calculations that was widely used for centuries.

These efforts did not, however, win everyone over to the new symbols. Those people who still favored the abacus stubbornly held on to the old ways and the old numerals, and were called *abacists*, while those who wanted to introduce the Indian numerals were *algorithmicists*.[10] The battle between the abacists and the algorithmicists continued for hundreds of years until the sixteenth century, long after the dawn of the Renaissance (which began in Italy during the thirteenth century.) Finally, a number of textbooks were printed in Germany called *Rechenbucher* that were so broadly accepted as the standard that the counting board and abacus lost out to pen and paper. Of course, by now the Europeans were producing much more paper, and scarcity was not as great a problem.

Why Europe resisted the Hindu–Arabic system is not fully explained by the shortage of paper in the first centuries. In fact, the failure to adopt a system with such clear advantages is a puzzlement. Both the counting board and abacus were based on position value, and the base-ten number system was common throughout Europe. Even the European languages formed the spoken numbers in a place-value manner. For example, in English we say "seven thousand, four hundred, nine." If we switch the words for "seven," "four," or "nine," we get a different number. The "thousand" and "hundred" are place holders. With the spoken languages, the abacus, and the base-ten number system all pushing the Europeans toward a modern place-value system, they still dragged their feet from the time it was introduced by Leonardo of Pisa until the end of the sixteenth century—a period of approximately four hundred years.

The adoption of both the zero and negative numbers paralleled the slow acceptance of the Hindu–Arabic number system. Some even considered the zero to be a creation of the devil.[11] The first time negative numbers appear in European equations was in 1484 in a work on arithmetic and algebra by Nicolas Chuquet, a Parisian doctor. One example of his use of a negative number is in the following equation: .4.[1] *egaux a* m̄.2.[0] which in our notation would be $4x = -2$. The small m with the bar above stood for a minus. It is remarkable that only five hundred years ago the symbols used in algebra were so different from those of today.

In 1489, just five years after Chuquet published his book, Johann Widman of Leipzig published a work on arithmetic that contained the use of our familiar plus and minus signs. Yet, it took some time for them to be accepted. It is believed that the plus and minus signs were first used by shippers marking boxes of freight as either overweight or underweight.

During the sixteenth century many algebraists struggled with a number system inherited from the Greeks that contained only positive numbers. The system was just too limited for their needs and each year more of them adopted the position that negative numbers were acceptable in computations, but pressure to deny that the solution of an equation could be negative still ran high. In their writings, these mathematicians showed the difficulty of accepting these strange numbers. Michael Stifel (ca. 1487–1567) used negative numbers in equations, but called them *"numeri absurdi."*[12]

Finally, the Flemish mathematician Albert Girard (1590–1633) in his work, *Invention nouvelle en l' algebre*, published in 1629, claimed that negative numbers had equal status with positive numbers. He accepted them as numbers used in equations and as solutions or roots of equations. He even went so far as to suggest that negative numbers are the opposites of positive numbers, which is how we are going to illustrate them on our number line. At long last, the negative numbers were taking their place in our concept of number.

Even by the end of the sixteenth century, acceptance of negative numbers and zero was not complete, for there were still a few stubborn souls who would not accept such strange entities. Even into the eighteenth century some textbooks did not provide for the multiplication of two negative numbers. An extreme example of this old-fashioned, Pythagorean thinking was embodied in the German mathematician Leopold Kronecker (1823–1891) who made the oft-quoted claim, "God made the integers; the rest is the work of man."[13] Kronecker actually believed that all the numbers except the natural numbers should be banished, for to base mathematics on nonexistent entities would result in contradictions. For a renowned mathematician to make this suggestion during the middle

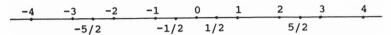

FIGURE 29. The number line with positive and negative whole numbers and fractions.

of the nineteenth century seems absurd, yet it illustrates that the influence of the Pythagoreans spanned two millennia to retard progress.

BACK TO THE NUMBER LINE

We are now ready to complete our number line with the numbers that have thus far been discovered, the natural numbers and fractions, the negative numbers, and zero. In Figure 29 we have located the negative numbers to the left of the zero in just the same manner as we did the positive numbers to the right of zero. Hence, for every positive number, fraction or integer, we have a corresponding negative number an equal distance from zero, but to the left. If we take all of these numbers that we have placed on the number line, that is, the positive and negative whole numbers (integers) and all the positive and negative fractions, then we have the *rational* numbers, that is, all numbers that can be written as one natural number divided by a second natural number. It took mankind until the seventh century A.D. to discover all the rational numbers. This was done by Brahmagupta in India, but his work and the work of his contemporaries were forgotten. The rational numbers were not accepted until the end of the sixteenth century in Europe, and even then stubborn "Pythagoreans" dragged their feet on this issue into the twentieth century.

However, the number system is still woefully incomplete, for the number line contains points that do not relate to any rational number. The story continues.

CHAPTER 8

Dealing with the Infinite

DIFFERENT KINDS OF INFINITY

Any meaningful discussion of numbers must include something on infinity. We have already thrown around the term in exploring the natural number sequence without any attempt to define its meaning. To understand what the natural numbers are is to have a sense for infinity, for understanding numbers implies infinity—that they go on "forever." Since mathematics begins with the study of numbers, we cannot appreciate either numbers or mathematics without tackling this strange and beautiful idea.

Certainly recognition of infinity is one of the greatest human accomplishments. Yet, it seems to cause much consternation and anxiety among people of all parts of society. Infinity is something you cannot quite wrap your mind around; it flows out, always leaving you behind. If one dwells too long on the infinite, then intellectual acrophobia sets in. However, others have embraced the concept of infinity, as if to believe in a completely finite world causes claustrophobia. Thus, humanity has oscillated between the fear of enclosure and the fear of falling into an endless void. How any single individual reacts to the concept of infinity may be more a function of his or her psychology than any underlining characteristic of the infinite.

The natural numbers suggest infinity; and it may have been their evolution that first suggested the concept to our ancestors. Prehistoric human beings, who failed to completely abstract numbers and separate them from the things being counted, probably believed, if they thought about it at all, that numbers were finite simply because all the objects they counted were finite. However, sometime in the dim past someone realized that numbers themselves could continue without end. How astonishing this must have seemed.

Now we must get over our amazement and try to develop a better understanding of the infinite. First, we must distinguish between the infinities that may exist in the universe—that is, physical infinities, and those that are purely intellectual—infinities of thought. What we are doing is distinguishing between the corporeal world and the world of ideas. The reason for this distinction is simple: to discover the different characteristics of each kind of infinity, the physical and the mental, requires different approaches. To investigate physical infinities we use empiricism; that is, we go out into the universe and look. This approach is at the heart of the modern scientific method. When our reasoning disagrees with our senses, we refashion our reasoning. This approach is distinctly non-Greek, for they thought that all real knowledge of the universe could be deduced from first principles, and they considered the information received by their senses as only opinion and not worthy of the label of truth.

The infinity of ideas, on the other hand, must be understood through deductive logic and not empiricism. Objects of thought act according to logic. Therefore, the Greek approach, that is, the application of deductive reasoning, is appropriate for investigating our mental ideas of infinity. One of the great historical struggles within mathematics was to construct a logic that would satisfactorily deal with the infinite, and it was not until the end of the nineteenth century that this was achieved.

A second distinction is between the infinitely large and the infinitely small. Most people, when thinking of the infinite, tend to think of bigness. But the concept goes in both directions, and the infinity of both the large and small have caused cosmological and logical problems. The question of bigness is called infinite magnitude while that of smallness is infinite divisibility or continuity. In reviewing the history of these problems we spend considerable time with the ancient Greeks because they discovered many difficulties associated with the infinite, and then got bogged down trying to extricate themselves. Most later thinkers adopted the Greek definitions and outlook, and fell into the same quagmires.

INFINITE MAGNITUDE

In terms of the physical or corporeal world, we have virtually no knowledge as to whether an infinite magnitude exists. Is the universe infinitely large? We do not know. "But how can the universe be finite?" says the claustrophobe. "It must go on forever, because, if we ever came to the end, what would be on the other side?"

The acrophobe responds, "Maybe the universe doubles back on itself. Maybe it is finite without being bounded."

But the claustrophobe will not let go. "That's impossible. You can't imagine an unbounded, finite universe, so it can't exist."

We can create similar hypothetical arguments concerning the infinite magnitude of time. How can the universe cease to exist? Must it go on forever? Did time always exist, or was there some beginning, as in the Big Bang Theory?

And so it goes, back and forth, each side using selected bits of logic to drive home their points. A unifying theme that often reappears in such arguments is whether one side or the other can "imagine" some concept. What is usually meant in these cases is the ability to visualize an idea internally. Can anyone visualize the universe continuing forever? Can anyone visualize the universe coming to an end with no boundary? The hidden premise here is that for something to have existence, human beings must be able to conceive of it, and in the most demanding case to "see" it with our "internal eye." How very peculiar. This particular argument has strange results, for it implies that before any animal was smart enough to conceive of a greater universe, such a universe did not exist. Then, when our ancestors evolved to a sufficient intelligence to look at the stars and wonder what was beyond, the universe sprang into existence as if by magic. In any case, the argument does not work. Corporeal things do not owe their existence to the fact human beings can conceive of them.

Another implication of the idea that thought sustains reality is the existence of an omniscient mind that constantly conceives of the universe and, thus, keeps it in existence. However, this does not imply that finite human beings cause corporeal objects to exist by our thinking. Therefore, whether we can or cannot conceive of something in the physical world has no bearing on its existence. For us, what exists, exists, and what does not, does not. We must be cautious to avoid this demand that we "see" with our inner eye what is being discussed and be willing to consider concepts that we cannot imagine in their totality.

The conclusion to all this philosophizing is that we cannot say whether any corporeal infinite magnitude exists. We just do not know. This is not all bad. It leaves the door open for us to speculate, and these speculations may permit us to conceive of reality in different, more interesting, ways.

The Greeks were the first to leave a substantial literary tradition regarding the infinite. The Greek word for infinity was *apeiron*, and, as we will see, infinity frequently worked as a causative agent in Greek cosmology. One of the first references to the concept of infinity comes from a fragment of a manuscript by Pherecydes of Syros (seventh or sixth century B.C.) who said, "Zas (Zeus) and Time existed always. . . ."[1] This demonstrates that Pherecydes believed the time dimension was infinite. Anaximander of Miletus (ca. 560 B.C.) also talked of infinite time, "This (essential nature, whatever it is, of the Non-Limited) is everlasting and ageless."[2]

It is Aristotle who gave the most comprehensive analysis of the infinite for the ancient Greeks (Figure 30). Book III of his *Physics* is almost entirely devoted to the idea of infinity. Aristotle criticized both Plato and the Pythagoreans for believing that the infinite was a "self-subsistent substance" rather than an attribute of something else. Aristotle did not accept that infinity was itself a substance rather than an attribute of some other existing entity. Aristotle also recognized the different applications of an infinite: infinite time, infinite magnitude, and infinite divisibility. Aristotle rejected infinite magnitude.

Our inquiry (as physicists) is limited to its special subject-matter, the objects of sense, and we have to ask whether there is or is not among *them* a body which is infinite in the direction of increase. We may begin with a dialectical argument and show as follows that there is no such thing.[3]

IDEAS OF THE INFINITELY LARGE

We have discussed infinite physical extension or the infinite in the physical world. What about abstract objects of thought, the world of ideas? Can abstract infinities exist? Now we are on firmer ground and can answer in the affirmative. Yes, we do have such ideas of the infinite. In fact, the best example, and the one which causes the least consternation

FIGURE 30. Aristotle, 384–322 B.C. (Photograph from Brown Brothers, Sterling, PA.)

among the acrophobics, is the idea of natural numbers. They go on forever. Now this is just where we want to be, for our goal is to understand numbers, and numbers are, considered as a collection, infinite.

Let's take a moment and discuss the ideas of the infinitely large found in natural numbers. People imagine whole numbers differently. As a boy I once thought everyone "saw" them the same way I did; that is, the number one in the center, two to its right at a little distance (so as not to confuse one and two with the number twelve), and then the number three to the right of two, and so forth. The natural numbers extended toward my right until they disappeared in a sort of haze. Then I asked my mother how she saw numbers. "They start at the top with the number one," she explained, "and kind of swirl down in a twisting form, getting larger." I was astounded. But, of course, it only makes sense that different people, with diverse backgrounds and training, would imagine numbers differently.

Most of us probably "picture" natural numbers "in our minds" as beginning with the number one and then progressing through a half dozen or more and finally disappearing out of our field of internal vision. We do not "see" the infinity of numbers. How do we know an infinity of numbers exists? It is because we can imagine what we think is the biggest number, and then add one to it to create an even bigger number. Hence, our biggest number was not the biggest after all. Therefore, we realize: There is no biggest number—they go on infinitely. This little argument, which we have all thought through, uses the common logical method known as the *indirect method of proof*, or *reductio ad absurdum*, which we pointed out previously as a favorite method of the Greeks.

Now we have our natural numbers, and they are infinite. But what do we mean by "infinite?" What is meant is that no natural number exists that is the largest, that is, the numbers are unbounded. This point is important, because in the sense we are using infinity here, infinity is not a number itself. It is, rather, a characteristic of the set of all natural numbers.

Through the ages, from the Greeks until the nineteenth century (and for some, even today) thinkers have been bothered by the infinity of numbers. They could not "see" all these numbers at once, so they felt that all those that could not be imagined must have some lesser existence than those small ones we use every day. This is the approach taken by the ancient Pythagoreans. Not liking the infinity of natural numbers, they made the first ten numbers the base to all else, with ten having a special sacredness. Philolaus of Tarentum was a Pythagorean of the second half of the fifth century B.C. who is reported to have influenced the philosophy of Plato. His words expressed the Pythagorean idea that ten is special in the generation of all numbers.

> One must study the activities and the essence of Number in accordance with the power existing in the Decad (*Ten-ness*); for it (the *Decad*) is great, complete, all-achieving, and the origin of divine and human life and its Leader; it shares . . . [sentence fragment lost here] The power also of the Decad. Without this, all things are unlimited, obscure and indiscernible.[4]

This belief implies that numbers greater than ten imitate the numbers one through ten and owe their existence to the repetition of those ten. Hence, larger numbers occur when the numbers one through ten repeat themselves a finite number of times. There is no infinity of numbers.

Plato also used the idea that some principle of ten-ness is working to limit infinite numbers. ". . . nor is the infinite in the direction of increase, for the parts number only up to the decad."[5]

Aristotle, too, rejected the existence of an infinite number of numbers. "Nor can number taken in abstraction be infinite, for number or that which has numbers is numerable."[6] Yet, he was clever enough to realize that the absolute denial of all infinities would lead to great awkwardness in his philosophy. To get around this awkwardness he invented a little sham. He said that the infinite was not "actual" but was only "potential." Hence, he denied full existence to the infinite while still trying to keep some idea of infinity alive.

> For generally the infinite has this mode of existence: one thing is always being taken after another, and each thing that is taken is always finite, but always different. . . . The infinite, then, exists in no other way, but in this way it does exist, potentially and by reduction.[7]

Aristotle wanted to admit that the natural numbers could always be added to greater and greater counts, but this, he claimed, was not infinity.

INFINITE DIVISIBILITY

The Greeks were obsessed with the question of whether space could be infinitely divisible, which subsequently caused them to be perplexed by motion and the continuity of space. The Eleatic School, founded by Parmenides of Elea (ca. 475 B.C.) was in opposition to the ideas of the Pythagoreans, who thought space consisted of line segments that were indivisible. The Eleatics not only showed the contradictions that arise from the idea of indivisible segments, they also showed the problems derived from the idea of space being made of infinitely many points.

Parmenides' metaphysics is a perfect example of the Greek desire to denounce mere sensation in favor of reason. Parmenides argued that since something cannot come out of nothing, then whatever is, simply is, and always has been. Being is *one* and unchanging. This led to the remarkable conclusion that motion must be an illusion, since real being cannot change. However, it was not Parmenides, with his remarkable philosophy of the *one*, who made a substantial impact on mathematics, but his student, Zeno of Elea (ca. 450 B.C.) (Figure 31). Zeno gave a number of arguments that demonstrated, according to Eleatic logic, that the world could not be a multiplicity of things or be changing, as opposed to the Pythagorean belief in numbers (monads) and magnitudes moving in space. His most famous argument is embedded in the story of Achilles and the footrace as retold by Aristotle.

> The second [argument] is the so-called "Achilles," and it amounts to this, that in a race the quickest runner can never overtake the slowest, since the pursuer must first reach the point whence the pursued started, so that the slower must always hold a lead. This argument is the same in principle as that which depends on bisection. . . .[8]

Zeno's argument on bisection simply says that motion is impossible because an object must first arrive at the halfway point to its final goal. Yet, before this, it must arrive at the halfway point on the way to the halfway point, and so on. Hence, it cannot move at all. What is assumed here by Zeno is that the distance between where the object begins and ends can be subdivided an infinite number of times, and the object cannot transverse an infinite number of points in a finite time. This notion of

FIGURE 31. Zeno of Elea, 489–? B.C. (Photograph from Culver Pictures, New York, N.Y.)

infinite divisibility is the central concept to the infinitely small. It helps us to see this concept in action if we use our number line and all the points we can identify on the line.

Figure 32 shows a portion of our number line between zero and one. We are going to change the subdivision process slightly, and move over ever smaller segments from zero to one. Beginning at zero we must first arrive at the point one-half. Next, we must cover the remaining distance and get to the point three-fourths. The next midway point is seven-eighths. Logically, this process could go on for an indefinite number of steps. We would never get to one in a finite number of steps because there

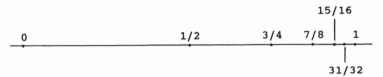

FIGURE 32. Traversing an infinite number of points to move from 0 to 1.

is always that little bit farther to go. To actually arrive at one (which we know happens in the physical world—i.e., we do move from place to place) we must transverse an infinite number of points. This bothered the Greeks, and it has bothered most mathematicians ever since. This may seem a silly point to quibble over, since we can all look about and see that things move. It is when we begin to ask ourselves just how things can move that Zeno's argument becomes meaningful. If points have no magnitude, then we can jam an infinite number of them into as small a space as we want. For an object to move, it must at one time or another occupy each and every one of the infinity of points. Yet, even that is not enough to get anywhere, because if points have no magnitude, then passing over an infinite number does not give us any movement whatsoever. It is this difficulty of handling motion and position that so bothered the Greeks and later philosophers.

The source of our uneasiness about infinite divisibility is our imagining an infinite number of points all jammed in between zero and one. In fact, the real "bulk" of these points in Figure 32 are jammed up against the one. How can they fit? But, remember, points have neither length nor height. They do not occupy any space, so neither do infinitely many points really occupy any space on the line between zero and one. However, our "feeling" is that the line is composed of points, and this fact presses us to reject the infinity of points. Of course, we have two questions here: Can physical space be infinitely divided, and can conceptual space be infinitely divided?

Numerous Greek philosophers, including the Pythagoreans, talked about unlimited elements or principles mixing with something else to compose the physical world. Plato, the father of idealism, is known primarily for his concept of perfect, unchanging forms. Yet, he too followed the Pythagorean tradition by using infinitely divisible space in his cosmology. In his *Timeaus* he speaks of the origin of the universe.

> And he made her [World Soul] out of . . . the indivisible and unchangeable [World of Ideas] and also out of that which is divisible and has to do with material bodies . . . which is space, and is eternal, and admits not of destruction and provides a home for all created things. . . .[9]

Plato also calls space a "receptacle of becoming" since it mixes in some mysterious way with the forms or ideas to generate the sensible world. The mixing must be mysterious, because Plato also maintains that the forms themselves never actually enter the physical world. The Pythagoreans mixed the *Limited* (number) with the *Unlimited* to create their universe. Numbers were like atomic particles and were indivisible, while the *Unlimited* was the chaos of space and was infinitely divisible. In some mystical manner, the *Unlimited* separated the numbers out of the first monad (the *One*) to generate corporeal things. Plato, influenced by the Pythagoreans, mixed the ideas with the receptacle of becoming (space) to generate the world. Both the *Unlimited* of the Pythagoreans and the receptacle of Plato had negative connotations. For both men, goodness resided within their opposite principles: the *Limited* and the ideas. For Plato, ideas (forms) were perfect and complete and not subject to division, while the receptacle was infinitely divisible. While Plato needed the principle of divisibility in his cosmology, it was not his favorite idea, and it was only in the *Timeaus* that he discussed it. While he begrudgingly accepted infinite divisibility and the infinite magnitude of time, he could not bring himself to accept infinite magnitude or infinite extension in space.

For Aristotle, infinite division was more difficult to handle than infinite magnitude. As with infinite magnitude, he said that the infinite division was a potential accomplishment and not an actual accomplishment. As Rudy Rucker, former Professor of Mathematics at Randolph–Macon Women's College, points out in his excellent book *Infinity and the Mind*,[10] Aristotle's distinction between actual and potential infinity is "doubtful" since he, in effect, invented a secondary reality to house his infinities. In the world of ideas, the natural numbers logically imply an infinite number of ideas. To call the infinity of them "potential" rather than "real" does nothing to increase our understanding of them.

THE INFINITELY LARGE BECOMES
THE INFINITELY SMALL

This reluctance to accept infinity on the part of the ancient Greeks is understandable because it is difficult to "see" all of the natural numbers in our mind's eye. As we try to visualize more and more numbers, they keep going out into space (toward the right for me), and no matter how hard we look, the numbers keep going—way out beyond Jupiter and Pluto and beyond the solar system. In order to assist us in visualizing the infinity of natural numbers we will use the number line in conjunction with a little projective geometry.* On the bottom of Figure 33 we have the number line with the natural numbers from one to ten and zero. Directly above zero is a circle. By drawing lines from zero and the natural numbers to the point representing the north pole on the circle, we can project all the natural numbers onto the circle. We associate each natural number with the point on the circle where the line and circle intersect. From the first numbers projected, we see that the natural numbers march upward on the circle toward the north pole.

However, the numbers never reach the north pole no matter how far toward the right we go on the number line and draw a line to the circle. In order for a natural number to exactly reach the north pole, its line would have to be parallel to the number line, and since all natural numbers begin on the number line, this cannot happen. Hence, the natural numbers are projected onto the circle and march ever closer to the point on top without ever reaching it. In fact, we can pretend that all natural numbers are projected onto the circle, and we get an infinity of points crowding together ever closer as we move farther up the circle. We can pretend to have a magnifying glass and look at the very top of the circle. We will see

*Projective geometry is that branch of mathematics concerned with those properties of geometrical figures that do not change when the figures are projected onto a different space.

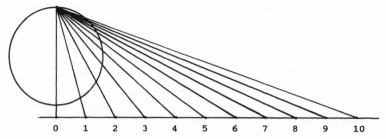

FIGURE 33. Projecting the infinite natural numbers onto a circle.

ever more number points marching toward the north pole point. Each is separated from its neighbors by ever smaller segments of the circle. This collection of points is similar to the infinity of points we got from Zeno's argument against motion.

What we have done, however, is perform some sleight-of-hand, a bit of flim-flam. For we have not constructed a way for us to "see" the infinity of points. This is impossible. What we have done is project the infinitely long number line onto a finite arc, allowing the points representing the numbers to come together on a finite curve. But at least we can now point to the circle and say that the infinity of natural numbers is represented by the set of points within a finite space. In short, we have traded the big infinity of the number line for the small infinity of divisibility.

STUCK IN A QUAGMIRE

The ancient Greeks could not reason themselves out of the various problems posed by the idea of infinity. Yet, the concept was, and is, essential to mathematics and the theory of numbers. Aristotle set the stage for handling the infinite: There were no actual, existing infinities but only potential infinities. Future mathematicians and philosophers tried to manage the infinite with this strange "potential" notion. The only exception was the God of Western religion. God must be infinite, and this was no problem since mortal man could not be expected to comprehend God.

The problems of the infinite, including infinite magnitudes, infinite divisibility, motion, and continuity, plagued Western thinkers until the nineteenth century, a period of twenty-two centuries after Aristotle's time. During this period many talented and intelligent men and women struggled to understand infinity, yet it remained an enigma. The following quotes demonstrate that the best minds of these two millennia were stuck in the mud.

Thomas Hobbes (1588–1679), an early British empiricist, dismissed the idea of the infinite because, he said, we cannot have infinite thoughts:

> Whatsoever we imagine is *finite*. Therefore, there is no idea or conception of anything we call *infinite*. No man can have in his mind an image of infinite magnitude nor conceive infinite swiftness, infinite time, or infinite force, or infinite power.[11]

This is using the argument previously outlined that if we cannot imagine something, it cannot exist. In this case, it relies on the notion that a thought of infinity would, itself, have to be infinite. Hobbes also demonstrated how confusing the ideas of motion and continuity were because of Zeno's paradoxes.

> Motion *is a continual relinquishing of one place, and acquiring of another* . . . I say a continual relinquishing, because no body, how little soever, can totally and at once go out of its former place into another, so, but that some part of it will be in a part of a place which is common to both, namely, to the relinquished and the acquired places.[12]

The Frenchman René Descartes (1596–1650) was both a philosopher and a mathematician of the first order. He is almost universally known for his famous statement "I think, therefore I am." He wanted to construct a rational science based on first principles, much like the mathematical model of geometry developed by Euclid. In his rationalism, Descartes shows himself to be still trapped in the method of Greek deductive science. In relation to the infinite he makes a most interesting argument for the existence of God. Descartes believed that since he was a finite being, he could only generate finite ideas. Since he had the idea of infinity, some infinite being must have given it to him. Hence, God, as an Infinite Being, must have put the idea of infinity into Descartes' mind.

And thus it is absolutely necessary to conclude, from all that I have before said, that God exists: for though the idea of substance be in my mind owing to this, that I myself am a substance, I should not, however, have the idea of an infinite substance, seeing I am a finite being, unless it were given me by some substance in reality infinite.[13]

Descartes was not unique in his quest to unravel infinity. Carl Friedrich Gauss (1777–1855) is considered by most mathematicians to be one of the three greatest mathematicians who ever lived. Yet, even Gauss could not free himself from a prejudice against the infinite. In a letter to a colleague he wrote:

As to your proof, I must protest most vehemently against your use of the infinite as something consummated, as this is never permitted in mathematics. The infinite is but a figure of speech . . .[14]

Hence, the best minds were still avoiding the infinite well into the nineteenth century. Even Kurt Gödel, one of the giants of twentieth century mathematics, could not rid himself of the uneasiness that an infinity of points could make up space. "According to this intuitive concept, summing up all the points, we still do not get the line; rather the points form some kind of scaffold on the line."[15]

EXHAUSTION AND LIMITS

Even though the Greeks, and most mathematicians following them, shied away from the infinite, certain problems required the use of infinity. For example, it was relatively easy for the Greeks, and the civilizations before them, to calculate the area of a square by squaring the length of one side. Rectangles were easy, and even triangles fell to simple computational rules. This is because these geometric forms all have sides that are straight. What about a circle? How can you calculate the area of a circle when it curves around everywhere?

For areas bounded by curves the Greeks developed the method of exhaustion, which pointed the way toward modern calculus. Consider Figure 34a, which is just a square that has been drawn inside of a circle. We can see that the area of the circle is greater than that of the square. Still, we can use the area of the square as a kind of first approximation to the area of the circle. Let's assume the diameter of the circle is one unit (one foot, centimeter, and so on). Now the question becomes: How good an estimate do we get by using the area of the square as the area of the circle? The area of the circle is πr^2 which is π times the square of the radius of the circle. In this case, the radius is 0.5 since the diameter is 1. Hence, we have approximately $(3.1416) \cdot (0.5)^2 = 0.7854$ for the circle's area. Since the diameter of the circle is the diagonal of the square, we can use the Pythagorean theorem to calculate the length a of one side and, from that, the square's area. The Pythagorean theorem states: $a^2 + a^2 = 1^2$ or $a^2 = 0.5$, which is the area of the square. Hence, the area of the square is 0.5—not very close to the figure of 0.7854.

We can do better. Notice in Figure 34a that four segments of the circle were not included in the area of the square. We can include most of the area of these segments by inserting triangles in them and then adding

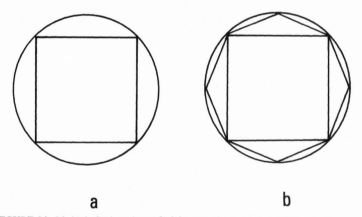

a   b

FIGURE 34. Method of exhaustion to find the area of a circle. In (a) the square is a poor estimate of the area of the circle. In (b) the square, plus the four triangles, is a much better estimate of the area.

the area of the triangles to the square. We do this in Figure 34b and determine the area of each triangle to be

$$\text{Area of triangle} = \tfrac{1}{2}(2r - a)\cdot(\tfrac{1}{2}a) = \tfrac{1}{2}a(r - \tfrac{1}{2}a)$$

If we multiply this by four we get

$$\text{Area of the four triangles} = a(2r - a) \text{ or}$$
$$\text{Area} = (1/\sqrt{2})(1 - 1/\sqrt{2}) = 0.2071$$

We now add this to the 0.5 (area of the square) and have for the new estimate = 0.7071. This number is much closer to 0.7854 than our first estimate of 0.5. Looking at Figure 34b, we see that there are now eight small segments that we have not included in our estimate. We can compute these areas and, adding them, get even closer to the true value of the area of the circle. However, no matter how many finite times we carry out this operation, we will never get the exact area of the circle, since there will always be small segments left out. The Greeks developed this method, which we call the *method of exhaustion*, to find the areas or volumes of curved figures. A rigorous treatment is given in Euclid's *Elements*, but it is known that Eudoxus developed the method earlier.

Euclid gave a proposition which states the exhaustion method.

> If from any magnitude there be subtracted a part not less than its half, and if from the remainder one again subtracts not less than its half, and if this process of subtraction is continued ultimately there will remain a magnitude less than any preassigned magnitude of the same kind.[16]

What this proposition says is that, if the error between our estimate and the real area is reduced by at least half of its value each time we construct a new set of triangles, then eventually we can get the error as small as we want. This idea is similar to Zeno's paradox of always traveling one-half the distance to the goal, never reaching it in a finite number of steps but still getting as close as we want. The idea of approaching a value as close as we desire is at the heart of modern limit theory, which in turn is the foundation of calculus. In addition, the theory of limits plays a key role in defining new kinds of numbers on the number line.

To understand limits we need some simple definitions. First we want to define a *sequence* of numbers as a special kind of collection of numbers.

> DEFINITION: A set of numbers: $A_1, A_2, A_3, \ldots A_n, \ldots$ forms a sequence of numbers if the numbers are well ordered, that is, if the subscripts are in the order of the natural numbers.

For example, we may have the sequence two, three, six, eight, which has a finite number of terms. However, the really interesting sequences have an infinite number of terms.

> DEFINITION: A sequence of numbers is infinite if each term of the sequence has a successor.

The easiest infinite sequence to visualize is the sequence of natural numbers or one, two, three, four, five, six, ... We indicate, as before, that this sequence is infinite with the three dots on the right. We have a simple rule to construct each successive term in the natural number sequence, that is, we just add the number one. If we were to ask ourselves how big the terms get, we know that there is no limit to their potential size.

The terms in some infinite sequences, however, never grow beyond a fixed value. For example, the individual terms in the sequence: $\frac{1}{2}, \frac{3}{4}, \frac{7}{8}, \frac{15}{16}$, . . . never grow larger than 1. We can see this at once when we consider the formula for the nth term which is $(2^n - 1)/2^n$. The formula shows that the numerator will always be less than the denominator. Hence, the sequence will never generate a term that is exactly equal to one or any term larger than one. Because the terms in our sequence do not grow infinite in size, we say the sequence is bounded—that is, the terms are limited as to how large they can grow. In fact, for the particular sequence above, we can say that the number one and all the numbers larger than one are bounds for the sequence. No term will ever be larger than these bounds.

We are now ready to ask one of the most fundamental questions of all mathematics. If we consider all the bounds for the terms of our sequence, is there a least (smallest) bound? Is there one bound smaller than all the others, yet still larger than any particular term in the sequence? For our sequence $\frac{1}{2}, \frac{3}{4}, \frac{7}{8}, \frac{15}{16}$, . . . the answer is yes, because 1 is that smallest bound. We call 1 the *limit* of the sequence. If we select any number smaller than 1 (but larger than $\frac{1}{2}$) then we can find a term of the sequence larger than that number, and still smaller than 1. This idea of a limit is the primary building block to higher mathematics. Before science and mathematics students can study higher mathematics in college, they must first take calculus. The first thing they learn in calculus is all about limits.

To fully appreciate the idea of a limit, we must understand what we mean when we say that the limit is the smallest bound. To say that a number, L, is a limit of a sequence means that the sequence will get ever closer to L without ever reaching it. In fact, if we pick a very small arbitrary number, ϵ, then we can look through our sequence until we find a term that will be closer to L by a value less than ϵ. We are now ready for a formal definition of a limit.

DEFINITION: The sequence of terms: $A_1, A_2, A_3, \ldots A_n, \ldots$
has a limit L if for any positive value ϵ these exists a
number N such that for all $n > N$ the absolute value of
$L - A_n < \epsilon$.

The above definition appears somewhat stilted and convoluted if you are not used to mathematical definitions. Yet, its underlying concept is easy to understand and this concept of limits is at the heart of all modern mathematical analysis. To restate the above definition in simpler terms, we say that the values of the numbers A_1, A_2, A_3, and so on, are getting ever closer to the value of L. For example, if we pick a very small value, say ϵ, then we can find a term in the sequence, say the Nth term, such that every term beyond the Nth is closer to L than $L - \epsilon$. We show this graphically in Figure 35. Here we have our sequence: $\frac{1}{2}, \frac{3}{4}, \frac{7}{8}, \frac{15}{16}, \ldots$ where each term is getting closer to 1. If you pick a small ϵ, say 1/10,000, then it is possible to find a term in the sequence such that every following term is closer to 1 than $(1 - 1/10,000)$. What would this term be? Taking successive values for 2^2 we see that $2^{14} = 16,384$. Therefore, the 14th term in the sequence is 16,383/16,384. This fraction is very close to 1, and is, in fact, closer than $1 - 1/10,000$ $(L - \epsilon)$, which is the fraction 9,999/10,000. It also follows that each successive term beyond the 14th term is going to be closer to 1 than our small error value of ϵ. We see now

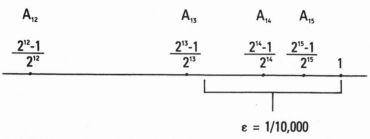

$$\varepsilon = 1/10{,}000$$

FIGURE 35. Finding a limit point. Even if we pick ϵ to be as small as 1/10,000, we find that the terms of the sequence beginning with A_{14} are a closer distance to 1 than ϵ.

that no matter how small an ϵ we pick, we can select a term in the sequence that will be closer than that to our limit of 1. This satisfies our desire to get as close to 1 as we want. With this sequence, we are getting closer to the number 1 (as in the Zeno paradox) without actually reaching it. In fact, no term in this sequence will ever reach or exceed 1. Therefore, the sequence is said to *converge* to 1.

> DEFINITION: A sequence that has a limit is convergent, otherwise it is divergent.

The reason for using the absolute value of $(L - \epsilon)$ in our definition of a limit is to accommodate approaching our limit from either direction. For example, the sequence $\frac{1}{2}, \frac{3}{4}, \frac{7}{8}, \frac{15}{16}, \dots$ approaches 1 from below (i.e., the values of all the terms are less than 1). The sequence $\frac{3}{2}, \frac{5}{4}, \frac{9}{8}, \frac{17}{16}, \dots$ is approaching 1 from above. It is possible to have a sequence that has limits but does not satisfy our definition. For example, the sequence $\frac{1}{2}, \frac{1}{4}, \frac{3}{4}, \frac{1}{8}$, $\frac{7}{8}, \frac{1}{16}, \frac{15}{16}, \dots$ has two limits (sometimes called cluster points) since alternating terms in the sequence converge to zero and 1. Yet, for our simple definition of limit, we will exclude such sequences.

We indicate the limit, L, of a sequence in the following way:

$$\lim_{n \to \infty} (A_n) = L$$

Here, the "lim" stands for "limit" and under the limit symbol we have $n \to \infty$, which means that the number of terms, n, increases without limit.

A_n is an abbreviation for the sequence. Hence, the limit of the sequence, as the number of terms goes to infinity, is L. If the sequence diverges then we show this as

$$\lim_{n \to \infty} (A_n) = \infty$$

which says that the terms keep increasing without limit.

One more example of a sequence from history will help illustrate a feature of mathematics that has enthralled mathematicians for millennia. Remember that we previously mentioned Leonardo of Pisa, who in 1202 helped introduce Hindu–Arabic numerals into Europe with his book, *Liber Abaci*. He was also called Fibonacci, and he discovered a fascinating number sequence that is named after him. It is simply 1, 1, 2, 3, 5, 8, 13, 21, ... Each term in the Fibonacci sequence is the sum of the previous two terms. This sequence diverges, of course, and we show this by equating the sequence to infinity:

$$\lim_{n \to \infty} (1, 1, 2, 3, 5, 8, 13, 21, \ldots) = \infty$$

However, we can form another sequence using successive terms from the Fibonacci sequence to make fractions, which gives us the sequence $\frac{1}{1}, \frac{2}{1}, \frac{3}{2}, \frac{5}{3}, \frac{8}{5}, \ldots$. Notice that each fraction is made from two successive Fibonacci terms with the larger term as numerator. This sequence converges nicely.

$$\lim_{n \to \infty} (\tfrac{1}{1}, \tfrac{2}{1}, \tfrac{3}{2}, \tfrac{5}{3}, \ldots) = (\sqrt{5} + 1)/2$$

Now the limit shown on the right of the equal sign might not mean much unless you were an ancient Greek mathematician or architect. This value, $(\sqrt{5} + 1)/2$, was known by the ancient Greeks as the Golden Mean or Golden Ratio, and was given its own symbol, ϕ. The Greeks considered this value a mean or ratio between two lengths because they did not have fractions. Hence, they thought of this value as simply the ratio between lengths $(\sqrt{5} + 1)$ and 2. Figure 36 shows a right triangle with the legs having lengths of 1 and 2. This, by the Pythagorean theorem, makes the hypotenuse equal to $\sqrt{5}$. Hence, if we divide the length of the hypotenuse $(\sqrt{5})$ plus the length of the short leg (1) by the length of the

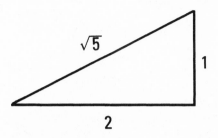

FIGURE 36. The Golden Mean or Golden Ratio is the ratio between 1 + √5 and 2 and is given the symbol φ.

long leg (2) we get (√5 + 1)/2 or the Golden Mean. The Greeks knew that a rectangle with a long side of √5 + 1 and a short side of 2 was pleasing to the eye, and they incorporated this ratio into much of their architecture, including the famous Parthenon in Athens. Later artists, including Leonardo da Vinci, used the Golden Mean in their work. The Pythagoreans used the pentagram as their symbol, and the pentagram incorporates several examples of the Golden Mean (see Figure 37).

With this example we can appreciate what frequently enthralls mathematicians—the interconnectedness of mathematics. Fibonacci's wonderful sequence has the Golden Mean as its limit, and the Golden Mean turns up in all kinds of unexpected geometrical constructions.

The idea of a limit is really an elegant and beautiful concept, for we are able to take an infinity of terms and relate them to something finite in the limit. Using this technique we are able to delve ever deeper into the mysteries of our number system to uncover startling and strange new numbers.

As a set of numbers, a sequence is sometimes awkward to integrate into a mathematical formula since a sequence, as a collection, is not equal to a single value in the normal arithmetic sense. An easy way to achieve this integration is to use another mathematical concept in place of the sequence—a series.

In our sequence $\frac{1}{2}, \frac{3}{4}, \frac{7}{8}, \frac{15}{16} \ldots$, we got closer to 1 by adding a small amount to each term to create the next term. We can rewrite this sequence as $\frac{1}{2}, \frac{1}{2} + \frac{1}{4}, \frac{1}{2} + \frac{1}{4} + \frac{1}{8}, \frac{1}{2} + \frac{1}{4} + \frac{1}{8} + \frac{1}{16}, \ldots$. Hence, we are adding $\frac{1}{2}$ of the value of the current term to get the next term. Each term in this sequence is a series of additions, or simply a series.

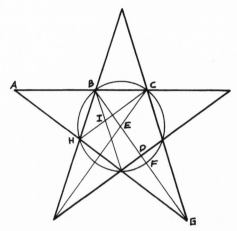

FIGURE 37. The pentagram, used as a symbol by the Pythagoreans, contains many Golden Means. ϕ = AB/BC = CH/BC = IC/HI = 2DE/EF = EG/2DE = $\sqrt{\text{EG/EF}}$.

DEFINITION: A series is the collective sum of a set of numbers.

Where a sequence is a collection of many different numbers, a series, as a sum, is just one number. Like the sequence, a series may have a finite or infinite number of terms.

DEFINITION: The series, $A_1 + A_2 + A_3 + \ldots A_n \ldots$ is infinite if there are infinitely many terms.

The following series is finite: $3 + 9 + 27$. We do not have to stop with just three terms, but can have an infinite number:

$$3 + 9 + 27 + 81 + \ldots$$

The above series increases in value beyond any limit we may try to impose. The series clearly diverges.

DEFINITION: If the sum of an infinite series is finite, then the series is convergent, otherwise it is divergent.

Now let's look at the series suggested by Zeno's paradox where it is necessary to move $\frac{1}{2}$ the distance, and then $\frac{1}{2}$ the remaining distance, and so on. Beginning with $\frac{1}{2}$, the series will look like $\frac{1}{2} + \frac{1}{4} + \frac{1}{8} + \frac{1}{16} + \ldots$. Mathematicians have developed a shorthand for representing a series. They use the Greek capital letter sigma Σ followed by a formula showing how to construct the series. The above series would be written as

$$\sum_{n=1}^{\infty} (1/2^n)$$

where n takes on the values 1, 2, 3, ... The $n = 1$ below the sigma indicates we begin the series with n equal to 1, and the infinity sign above the sigma means we continue through all the natural numbers. The sigma indicates we are to sum all the terms that will result in a single number. On occasion, we abbreviate the above series as just $\Sigma (1/2^n)$, or in the general case as ΣA_n.

DEFINITION: The series, $A_1 + A_2 + A_3 + \ldots A_n \ldots$ can be written as

$$\sum_{n=1}^{\infty} (A_n)$$

where A_n is the nth term.

Series, just like sequences, can have limits. If we imagine that we can quickly add up the infinite number of terms of the series, $\Sigma (1/2^n)$, what would its value be? It would be 1. That is, 1 is the limit of the series. But this way of talking is unsatisfactory to mathematicians, for they do not like to talk of "quickly adding an infinite" number of terms. This is a little too vague for them. Therefore, we approach the definition of a limit of a series without using the term "infinite." We say that a series, ΣA_n, has a limit if $\Sigma A_n = L$.

DEFINITION: The series ΣA_n has a limit if there exists a number L such that $\Sigma A_n = L$.

To appreciate that L is a limit we must realize the L is the exact value or number that is equal to the sum of all the terms in our infinite series. Frequently we can write a series so that the nth term shows how the series is constructed. For example, for the series $\frac{1}{2} + \frac{1}{4} + \frac{1}{8}$ . . . we can write $\Sigma\,(1/2^n)$. This is a very convenient way to represent the series, and it shows the reader how the series is to be constructed. Many converging infinite series are known, and some of them have remarkable limits.

$$\frac{1}{2} + \frac{1}{3} + \frac{1}{5} + \frac{1}{7} + \frac{1}{11} \ldots = \Sigma\,(1/\text{prime}) = \infty$$

$$1/1^2 + 1/2^2 + 1/3^2 + 1/4^2 + \ldots = \Sigma\,(1/n^2) = \pi^2/6$$

$$1 - \frac{1}{3} + \frac{1}{5} - \frac{1}{7} + \frac{1}{9} - \ldots = \pi/4$$

The first series is just the sum of the inverses of all prime numbers. This series, like the series of the inverses of all natural numbers, diverges. The second series is the sum of the inverses of the squares of consecutive numbers and its limit was discovered by the great mathematician Leonhard Euler (1707–1783). Notice that π is a term in the limit. How could such a simple series be related to the ratio of a circle's diameter and its circumference? The third series is called an alternating series since the signs of the terms alternate. It, too, has a limit that involves π. In fact, π turns up in all kinds of unexpected places. This again demonstrates the wonderful connectedness of mathematics.

Given the ideas for series and sequences and their limits, we can rationally discuss the attributes of continuous change. In higher mathematics, limits permit us to study continuous difference rather than just the incremental differences suggested by the natural numbers. Hence, we can unlock the mathematical secrets to curves, accelerations, and movements. Modern mathematics and science could never have been developed if we had not discovered limits. Understanding series and sequences with their associated limits prepares us for our next great expansion of the number system—the irrational numbers.

CHAPTER 9

Dedekind's Cut
Irrational Numbers

We have studied thus far the various kinds of numbers that evolved from the natural numbers: the fractions, negative numbers, and zero. All these numbers together are called *rational numbers*. Such numbers are familiar to everyone in normal daily activities, from balancing checkbooks to playing blackjack. We are now at the point of learning about kinds of numbers most people are either unaware of, or only encountered years ago in a high school algebra class. Yet, these new numbers are amazing objects that can dazzle and puzzle mathematicians and laymen alike.

Using algebraic equations as a format, we can see why mathematicians since the Greeks have been frustrated in finding all the solutions to problems they felt should be available. The negative numbers solved such elementary equations as $x + 5 = 0$. With the acceptance of negative numbers, the answer is simply -5. But what about $x^2 - 2 = 0$? In other words, what number, when multiplied by itself, is equal to 2? Since the time of the Pythagoreans, mathematicians knew that the square root of 2 could not be represented by any fraction, and hence was not a rational number. Therefore, the set of all rational numbers was simply inadequate to account for all the desired solutions to algebraic equations. Did other, irrational, numbers exist that would solve such equations? Let's consider our number line again. Figure 38 shows the number line between zero and 2. We have taken the hypotenuse of a right triangle with each leg

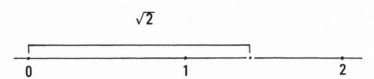

FIGURE 38. The point on the number line below the end of the length equal to $\sqrt{2}$ is not a point associated with any rational number (i.e., a number that is the ratio between two whole numbers).

having a length of 1 and placed this hypotenuse over the number line. The left end of the line segment is over the zero. To what point on the line does the other end correspond? We know the geometric length of this line segment is $\sqrt{2}$ from the Pythagorean theorem. If we drop a line segment from the end of the $\sqrt{2}$ length onto the number line, the point where it intersects the number line will not represent a rational number. It is this point that we wish to associate with $\sqrt{2}$ just as we associated another point with the number 2. We will call this new number an irrational number. Just what kind of number is it that is associated with this point on the number line?

Since antiquity, $\sqrt{2}$ was identified with one of these new irrational numbers. Do other examples of these irrational numbers exist? Since we have already proven that $\sqrt{2}$ is irrational, we can easily prove that the values for $\sqrt{3}$, $\sqrt{5}$, $\sqrt{6}$, and the square roots of all natural numbers which are not themselves squares, are also irrational. Hence, $\sqrt{4}$, $\sqrt{9}$, $\sqrt{16}$ are rational numbers but all roots of nonsquares are irrational. This means that there must be an infinite number of irrationals. We can form many additional irrational numbers. For example, the number whose value is equal to $(2 + \sqrt{2})$ must also be irrational. We can prove this by assuming that $(2 + \sqrt{2})$ is a rational number and showing this assumption leads to a contradiction. If $(2 + \sqrt{2})$ is rational then there exist two natural numbers, a and b, such that $2 + \sqrt{2} = a/b$. Now we simply subtract 2 from both sides of the equation to get $\sqrt{2} = a/b - 2$. However, the right side of this equation is obviously a rational number since both a/b and 2 are rational. This implies that $\sqrt{2}$ is rational, and we already know that it is not. This contradiction proves that $(2 + \sqrt{2})$ must be irrational. Of course, we can form many more combinations of rational and irrational numbers and use the same logic to prove the results are irrational numbers, which demonstrates that an infinite combination of rationals and irrationals results in even more irrationals.

So far we have only considered square roots of natural numbers, but there are many more possibilities for irrational numbers. Consider the equation: $x^3 - 5 = 0$. Solving for X we have $x = \sqrt[3]{5}$, which means a number which, when multiplied by itself twice, is equal to five. This is a cube root, and it is also irrational. In like fashion, we can define other nth roots where n is a natural number, and almost all of these will be irrational. In fact, we can give a little theorem which tells us which roots of whole numbers we can expect to be irrational.

THEOREM: If n and a are natural numbers, then the nth root of a is rational if and only if a is the nth power of another natural number; otherwise it is irrational.

For example, is the square root of 9 irrational or rational? Since 9 is the square of 3, which is a natural number, the square root of 9 is rational. Consider the cube root of 9. Is this number rational? We know the cube root of 9 is not a natural number because 2 cubed is 8 while 3 cubed is 27. Therefore, the cube root of 9 must be a number between 2 and 3. Hence, whatever it is, it is not a natural number. Since it is not a natural number, then by the above theorem, it must be irrational. In a similar fashion, the cube root of 8 is rational because 8 is the cube of 2 (a natural number), but the square root of 8 is irrational since it is not a natural number.

In each case where we have an irrational number we can construct a geometric magnitude corresponding to that number and lay it on our number line, with one end positioned above the zero. Where the other end falls, we have a point on the number line that does not correspond to a rational number. Therefore, the number line must be full of these "irrational" points.

We call these new numbers irrational numbers but this is only a name. To get a better understanding of irrational numbers we must look at two characteristics of the rational numbers. First, the rational numbers are *simply ordered*.

DEFINITION: The numbers in a set are simply ordered if two
conditions hold:

  (1) For any two numbers, A and B, in the set, one and
      only one of the following holds: $A > B$, or $B > A$, or
      $A = B$.
  (2) If A, B, and C are numbers in the set, and if $A > B$
      and $B > C$, then $A > C$.[1]

The above two conditions seem self-evident for numbers because we
are so used to using them when we manipulate numbers. It seems juvenile
to say that any two whole numbers must be equal or that one must be
greater than the other. Yet, it was not clear that these strange irrational
numbers had this kind of order. When we look at rational numbers we
"see" their order because of the number system we use. Hence, by
inspection, we see that 27,991 is larger than 3,990 and that 0.199987 is
smaller than 0.1999998. That is, the number symbol gives us the
number's order. But is $(\sqrt{2})\cdot(\sqrt{3}) = (\sqrt{6})$? If not, which is bigger?
From this example, we realize that the current number system does not
always reveal the order of irrational numbers, and we might ask: Are they
simply ordered as are the rational numbers?

The second question has to do with closure. We defined closure
earlier. If a set is closed under an operation, then each time we use that
operation on the elements of the set we get a result that is also in the set.
For example, the natural numbers are closed under both addition and
multiplication. This tells us that whenever we add two natural numbers,
we always get a natural number for an answer. Again, when we multiply
any two natural numbers we get another natural number.

But the reverse operations of subtraction and division are not closed
for the natural numbers. For example, $7 - 12 = -5$. Here we subtracted
two natural numbers and got a negative number. When we divide two
natural numbers we almost always get a proper fraction, not a whole
number. On the other hand, all the rational numbers (natural, negative,
fractions, and zero) are closed under all four of the primary operations,
with the single exception of dividing by zero. Hence, we might say that
the rational numbers are virtually closed. We can see why closure is an
important property. It means that when we use the four operations of
arithmetic, we know what we get back—another rational number.

What happens when we combine the irrational numbers with the rational numbers? Do we still have closure? We can see at once that the irrational numbers by themselves are not closed for we have $(\sqrt{2}) \cdot (\sqrt{2})$ = 2. Hence, we have a case where the product of two irrational numbers is rational. The larger question is whether we lose closure when we include the irrationals with the rational numbers. If we start using the four operations of arithmetic on irrational numbers, will we begin to get strange nonnumber entities as answers?

The general discomfort that mathematicians felt about irrationals went beyond the questions of closure and order. Modern mathematical analysis is based on calculus, and calculus is founded on the theory of limits. But the original theory of limits used ideas from geometry, such as a sequence of points approaching a limiting point. But could all such limiting points be considered to represent legitimate numbers? Would the whole house of cards holding up higher mathematics fall down because the theory of irrational numbers was not firmly based on the idea of rational numbers?

EUDOXUS'.RATIOS

Before we proceed with the modern treatment of irrational numbers we should take a moment and review how the ancient Greeks tried to handle them. Earlier we showed how the discovery of $\sqrt{2}$ upset the Greek mathematicians because they found a line segment that could not be represented as the ratio of natural numbers (the only numbers they accepted). This caused a separation of geometry and algebra because all the objects of geometry could not be represented in algebra. But even with this separation, problems remained for geometry. The geometry of Euclid used proofs relying on the idea that geometric magnitudes (lengths, area, volumes) could be compared to each other. Yet, some of these magnitudes were incommensurable lengths. Could such magnitudes be compared to each other? The Greek geometers needed a theory of incommensurable magnitudes.

Eudoxus of Cnidus (408–355 B.C.) was a student of Plato who became one of the greatest Greek mathematicians. He was also a doctor and astronomer and is credited with the theory of concentric spheres to explain the apparent motions of the heavenly bodies. Eudoxus proposed a theory of proportion to explain how incommensurable magnitudes could be used in geometry. His theory has been preserved for us in the fifth book of Euclid's *Elements*. The theory, as stated, was very oblique and difficult. It was pondered by mathematicians until it was superseded in the nineteenth century. His definition of proportions in Euclid's work exemplifies the struggle taking place in the Greek mind to get a handle on this problem.

> Magnitudes are said to be in the same ratio, the first to the second and the third to the fourth, when, if any equimultiples whatever be taken of the first and third, and any equimultiples whatever of the second and fourth, the former equimultiples alike exceed, are alike equal to, or alike fall short of, the latter equimultiples taken in corresponding order.[2]

What could such an inscrutable statement possibly mean? It seems that Eudoxus (through Euclid) must have sat up nights trying to write something that no one could comprehend. To understand this statement we must remember two things about Greek mathematics. First, Eudoxus was not talking about numbers, but magnitudes. The two were not the same and could not be related to each other. Second, the Greeks did not have fractions, so they spoke of the ratios of numbers and ratios of magnitudes. Hence, our fraction $\frac{2}{3}$ was for them the ratio 2:3. For their geometry, they also needed to talk about ratios, not of numbers, but of geometric magnitudes. For example, they knew that the ratio of the areas of two circles is equal to the ratio of the squares of the diameters of the circles. We can show this as

$$\text{(area of circle A):(area of circle B)} =$$
$$\text{(radius of circle A)}^2\text{:(radius of circle B)}^2$$

The Greeks had to be sure that when these ratios of magnitudes involved incommensurable lengths, the order relationships held. In other words, would their geometric proofs be valid when such proofs involved ratios of incommensurable lengths? The definition developed by Eudoxus was an attempt to guarantee that they would. The magnitudes in the ratios have the following labels: first:second = third:fourth.

Eudoxus said that the first and second magnitudes have the same ratio as the third and fourth if, when we multiply the first and third by the same magnitude, and multiply the second and fourth both by another magnitude, then whatever order we get between first and second will be preserved between the third and fourth.

This explanation, simple as it is, can be rather confusing. An example will clarify the matter. We will assign the following lengths to the four magnitudes: $3:6 = 7:14$. From this we get the following inequalities: $3 < 6$ and $7 < 14$. Eudoxus says that if we multiply both 3 and 7 by any magnitude A and we multiply both 6 and 14 by any magnitude B, then the order relationship between the altered 3 and 6 will be the same as between the altered 7 and 14. Let $A = 5$ and $B = 2$. Multiplying we get

$$A \cdot 3 : B \cdot 6 = A \cdot 7 : B \cdot 14 \text{ or } 15:12 = 35:28.$$

Now clearly $15 > 12$ and $35 > 28$. Hence, multiplying by 5 and 2 preserved the order of the two ratios. Eudoxus' definition says that for two ratios to be equal, all values of A and B will preserve the order between the corresponding magnitudes. This gave Greek geometry the definition of magnitudes of ratios it needed to carry out the various proofs relying on proportion. However, magnitudes are not numbers, and the requirement that all values of A and B satisfy the definition introduced, through the back door, the notion of infinity. While Eudoxus' work satisfied the needs of geometers, it was still inadequate to meet the needs of arithmetic and establish irrational numbers on a sound theoretical base.

Before we give the final solution to irrationals, we must give credit to those mathematicians who were willing to accept irrationals among the family of numbers. The Hindus in general and Brahmagupta in particular considered irrational roots as legitimate solutions to equations. Hence, they had no qualms about accepting $\sqrt{2}$ or $\sqrt{3}$ as a solution to a problem. Another famous mathematician, known in the West as a poet, was Omar Khayyam (ca. 1050–1123). While the Greeks had used geometric methods to solve equations that treated the solutions as lengths, he treated the solutions as numbers. He also worked on the problem of defining irrationals.

During most of the Middle Ages, European mathematicians allowed only the natural numbers and fractions as solutions to equations. However, the symbolic nature of algebra progressed, and eventually the old rhetorical method was abandoned. When words were replaced by symbols, the negative and irrational numbers (represented by the square roots of integers) began to creep in as solutions. To say, in rhetorical form, that seven geese less eleven geese are a minus four geese strains the imagination, for what could a "minus four geese" possibly mean? But to use symbolic notation hides this difficulty for we have $7 - 11 = -4$. With this abstraction of algebra, mathematicians began to use negative and irrational numbers without any acceptable definition as to just what they were.

The three mathematicians most often considered the greatest are Archimedes, Sir Isaac Newton, and Carl Friedrich Gauss. It was Carl Gauss (1777–1855), born into a laborer's family and considered a child prodigy, who helped push irrationals into acceptance (Figure 39). As a student he produced one of his masterpieces, *Disquisitiones Arithmeticae*, for his doctoral dissertation. Yet, the manuscript was held up for three years in publication. In the meantime, he wrote another little dissertation to satisfy his degree requirement, which was published in 1799 when he was twenty-two. In this second paper Gauss proved what has become "The Fundamental Theorem of Algebra." What theorem could possibly merit such a distinction? The theorem says that every polynomial equation* with one unknown has at least one solution or root. A corollary to this theorem is that there are as many solutions as the highest exponent on the unknown. Therefore, the polynomial equation

$$A_0X^n + A_1X^{n-1} + \ldots + A_{n-3}X^3 + A_{n-2}X^2 + A_{n-1}X + A_n = 0$$

has exactly n solutions. In some cases we get identical solutions as in the equation $x^2 - 10x + 25 = 0$. If we factor this equation we get $(x - 5)(x - 5) = 0$ or $x = 5$ for both solutions. The important point for our discussion is that the theorem says that an equation such as $x^2 - 2 = 0$ has two solutions. These solutions are, in fact, $\sqrt{2}$ and $-\sqrt{2}$, both irrational numbers. Therefore, with the proof of this fundamental theorem, Gauss was forced to accept irrational numbers as solutions to equations. As we

---

*A polynomial equation is an equation where the left side consists of one or more unknowns (generally x's, y's, etc.) raised to powers and multiplied by numbers called coefficients. The right side of an equation is generally zero. For example, $5x^2 + 3x - 4 = 0$ is a polynomial with one unknown (x) and coefficients of 5, 3, and -4.

FIGURE 39. Carl Friedrich Gauss, 1777–1855. (Photograph from Brown Brothers, Sterling, PA.)

will see later, Gauss also set the stage for acceptance of another entirely new class of numbers, the complex numbers.

With Gauss forcing the issue at the beginning of the nineteenth century, the search was on for a meaningful definition of the irrationals.

THE WONDERFUL DEDEKIND *SCHNITT*

We are especially fortunate when it comes to the modern definition of irrational numbers. All of our rational numbers are buried so far into the past that we cannot say when someone first counted on his or her fingers and said, "Ah ha! Natural numbers!" Nor can we read the thoughts of the mathematicians who discovered the fractions or the negative numbers. But the logical foundation to the irrational numbers is recent enough that we can purchase and read our own copy of the wonderful text. The definition of irrational numbers is contained in a thin manuscript by Richard Dedekind (1831–1916), a German mathematician of the nineteenth century, and is entitled *Continuity and Irrational Numbers*.[3] Anyone picking up this work can sit down and read it in its entirety in a few minutes. It is concise, yet powerful.

Richard Dedekind was born in Brunswick, Germany. At the age of seventeen he entered college and at twenty-one received his doctorate from the University of Göttingen in 1852, accepting his degree from the already famous Carl Gauss.

Although Dedekind discovered his theory in 1858, he did not publish it until 1872. In the beginning of *Continuity and Irrational Numbers* he tells us that he was dissatisfied with the idea of continuity used in calculus, which relied on notions borrowed from geometry. After considering the problem he came to his solution on November 24, 1858.

The first thing Dedekind did was to define the set of all rational numbers as the *system R*. He went on to say that the following laws held for the numbers in *R*:

I. If $a > b$, and $b > c$, then $a > c$. Whenever a, c are two different (or unequal) numbers, and b is greater than the one and less than the other, we shall, without hesitation because of the suggestion of geometric ideas, express this briefly by saying: b lies between the two numbers a, c.

II. If a, c are two different numbers, there are infinitely many different numbers lying between a, c.

III. If a is any definite number, then all numbers of the system R fall into two classes, A_1 and A_2, each of which contains infinitely many individuals; the first class A_1 comprises all numbers a_1 that are $<a$, the second class A_2 comprises all numbers a_2 that are $>a$; the number a itself may be assigned at pleasure to the first or second class, being respectively the greatest number of the first class or the least of the second. In every case the separation of the system R into the two classes A_1, A_2 is such that every number of the first class A_1 is less than every number of the second class A_2.[4]

It would be difficult to state Dedekind's idea more clearly. The first law we have already encountered; it says that the rational numbers are ordered. The second law says there are infinitely many numbers between any two rational numbers. This is the ancient Greek idea of infinite divisibility. The third law is the real meat of his idea. It says that each rational number divides all the rational numbers into two classes where all the numbers in the first class are smaller than those in the second class. It then states that the rational number used to make this separation of numbers can be included in either class. It can be the largest number in class A_1 or the smallest number in class A_2.

So far I have quoted Dedekind talking only of rational numbers. When he was ready to define the irrationals, he simply dropped the requirement from his third law that all separations of the rational numbers be defined by a rational number. Was he correct in doing this? He stated:

> If now any separation of the system R into two classes A_1, A_2 is given which possesses only this characteristic property that every number a_1 in A_1 is less than every number a_2 in A_2, then for brevity we shall call such a separation a *cut* [Schnitt] and designate it by (A_1, A_2).[5]

We may ask why Dedekind felt it necessary to define a new term, cut, instead of using the idea of a point dividing all the rational numbers on the number line into two classes. We must remember that he wanted to divorce the idea of irrational numbers from geometric notions, such as points and lines. In the above definition of cut he has achieved this, for he has defined a cut in terms of two classes of numbers where every number of the first class is less than every number of the second class. He cleverly avoided mentioning lines or points.

He defined his cut without saying how the cut is to be carried out. We already know we can produce such a cut using any rational number. Now the question is whether a cut can be made without a rational number. Dedekind answered this in the following:

> But it is easy to show that there exist infinitely many cuts not produced by rational numbers. The following example suggests itself most readily.
>
> Let D be a positive integer but not the square of an integer, then there exists a positive integer λ such that

$$\lambda^2 < D < (\lambda + 1)^2$$

If we assign to the second class A_2 every positive rational number a_2 whose square is $>D$, to the first class A_1 all other rational numbers a_1, this separation forms a cut (A_1, A_2), i.e., every number a_1 is less than every number a_2. . . . This cut is produced by no rational number.[6]

What is going on here? First, he points out that for every integer D which is not itself a square of a natural number, that D is bracketed between two other squares whose square roots are whole numbers that differ by 1. For example, 5 is bracketed by 4 and 9, and 4 is 2^2, while 9 is 3^2. He then chooses his cut so that A_2 is the set of all numbers *whose squares are greater than 5*. By this very clever approach, Dedekind has made a cut that defines $\sqrt{5}$, which is an irrational number, yet he uses only rational numbers in his definition of this cut. Thus he avoids using irrational numbers to define irrational numbers.

For a second example, we can define a cut for the irrational number $\sqrt{2}$ in the following way: let A_2 contain all the rational numbers whose squares are greater than 2. All other rational numbers will be in A_1. This takes a little thinking, but will soon pop clearly into the mind, just as it must have done for Dedekind back on November 24, 1858. Look at Figure 40, which shows the location on the number line of both $\sqrt{2}$ and 2. Now imagine any rational number that is larger than $\sqrt{2}$ (which would be to the right of $\sqrt{2}$). We know the square of that rational number must be larger than 2.

By this wonderful device of using only the rational numbers to create cuts that separate the rational numbers, Dedekind has defined the irrational numbers.

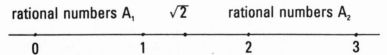

rational numbers A_1 $\sqrt{2}$ rational numbers A_2

0   1   2   3

FIGURE 40. A Dedekind cut. All the rational numbers whose squares are less than 2 are on the number line left of $\sqrt{2}$ and in A_1. All the rational numbers whose squares are greater than 2 are on the number line right of $\sqrt{2}$ and in A_2. These two sets of rational numbers, A_1 and A_2, define the irrational number $\sqrt{2}$.

Whenever, then, we have to do with a cut (A_1, A_2) produced by no rational number, we create a new, an *irrational* number a, which we regard as completely defined by this cut (A_1, A_2); we shall say that the number a corresponds to this cut, or that it produces this cut. From now on, therefore, to every definite cut there corresponds a definite rational or irrational number, and we regard two numbers as *different* or *unequal* always and only when they correspond to essentially different cuts.[7]

The set of numbers that is the combination of all rational and irrational numbers is called the *real* number system. By defining all real numbers by cuts, that is, sets of infinitely many rational numbers, Dedekind essentially defined numbers with infinite sets. Hence, our number system depends on the idea of infinity for a secure foundation. The ancient Greeks never could have accepted this notion as representing the underpinnings of mathematics.

Not only does Dedekind define the irrational numbers, but he goes on to define the operation of addition for all real numbers with the use of cuts. Once this is accomplished, it is an easy task to define the other three operations in arithmetic for real numbers. His definitions allow him to assert that when real numbers (rational or irrational) are manipulated with the four operations of arithmetic, the result will always be another real number (excluding division by zero). This gives him closure of the real number system, and allows him to assert the simple equality: $\sqrt{2} \cdot \sqrt{3} = \sqrt{6}$. Now that Dedekind has secured the irrational numbers for us, we can ask: Exactly what are they?

FUN WITH THE DECIMAL SYSTEM

The invention of the decimal system, which uses a dot to separate the integer part of a number from its fractional part, was a great boon to calculating. Fractions were messy to add and subtract, and some were difficult to compare. For example, which is larger, $\frac{21}{73}$ or $\frac{143}{517}$? It is hard to say. But, when we look at the first four digits of their decimal representations we can tell at once, for $\frac{21}{73} = 0.2877$ while $\frac{143}{517} = 0.2766$. From the decimals we see at once that $\frac{21}{73}$ is the larger.

Decimal fractions were introduced early in the Renaissance. In

1492, Francesco Pellos (fl. 1450–1500) published his work *Compendio de lo abaco*, which included the use of a dot to indicate a fraction having a denominator that is a power of ten.[8] However, the ancient Babylonians had used sexagesimal fractions based on sixty. The practice of using sexagesimal fractions was adopted as the standard for astronomy in the Western world and was used in general calculations during the Dark Ages. For some time after Pellos, the sexagesimal fractions held on. Finally, the French mathematician, François Viète (1540–1603), argued for the decimal system, and then John Napier (1550–1617) made it popular by introducing the decimal point to separate the integer part of a number from the fractional part.[9]

The modern decimal system is a shorthand for representing fractions (or fractional parts of numbers) as the sum of fractions whose denominators are increasing powers of 10. Hence, for 834.572 we have

$$834.572 = 8{\cdot}10^2 + 3{\cdot}10^1 + 4{\cdot}10^0 + 5/10^1 + 7/10^2 + 2/10^3$$

or

$$834.572 = 800 + 30 + 4 + \tfrac{5}{10} + \tfrac{7}{100} + \tfrac{2}{1000}$$

The utility of the decimal system is apparent at once when we realize how compact a space we can write a number in, and the ease of manipulating the number in arithmetic operations.

However, there is a peculiarity with decimal representation. To represent a fraction as a decimal we generally divide (using long division) the fraction's denominator (bottom) into the numerator (top). For example, the fraction $\frac{2}{5}$ is 2 divided by 5, which is 0.4. This decimal is called a terminating decimal because the digits to the right of the decimal point are finite—that is, they come to a natural end.

Now let's try to change the fraction $\frac{1}{3}$ into a decimal. We divide 3 into 1 in the following way: 3 does not divide into 1 directly, so we write down a decimal point and attach a zero to the 1 to make it 10. Now we divide 3 into 10 to get an answer of 3 with 1 as a remainder. We write the 3 to the right of the decimal point to get 0.3, but still have the 1 as a remainder. To this remainder of 1 we attach another zero to get 10 and repeat the process all over again. Yet, every time we divide by 3 we get 3 with a remainder of 1 and the process never ends. Therefore, to show the exact decimal value for $\frac{1}{3}$ we would have to show an infinite string of 3s. To get around this nuisance we show three dots trailing to the right of the last 3 to indicate that the 3s go on forever. Hence, $\frac{1}{3} = 0.333$ . . . and we understand that a finite number of 3s is always an approximation of $\frac{1}{3}$. Some texts show this repetition with a dot over the last 3 or $0.33\dot{3}$ which means that the figure under the dot is repeated. This kind of decimal is called an infinite, repeating decimal. Many fractions result in infinite, repeating decimals, and many of these have more than one repeating digit. Consider the fraction $\frac{3}{11}$ which yields the decimal 0.272727. We repeated the cycle of 27 three times, but the number of repetitions is up to the individual. We could have simply written it as $0.\dot{2}\dot{7}$.

For fractions we can say: Every fraction can be represented as either a terminating decimal or an infinite, repeating decimal. Conversely, every terminating decimal and infinite, repeating decimal can be represented as a fraction. To change from a decimal back into a fraction is not difficult. First, we handle the terminating decimal. Suppose our decimal is 1.028. We count the number of digits to the right of the decimal point. There are three. We now show the decimal as a numerator with 1 as the denominator. We multiply both the numerator and denominator by 10^3. The exponent of 10 is three because we counted three digits to the right of the decimal point. Hence, we get

$$1.028 = 1.028/1 = (1.028)\cdot(1000)/1\cdot1000 = \tfrac{1028}{1000}$$

We now have a fraction and can reduce it to its lowest terms by canceling the common factors from the numerator and denominator, or $\frac{1028}{1000} = \frac{257}{250}$. Now we know that the decimal 1.028 equals the fraction $\frac{257}{250}$.

Changing infinite, repeating decimals to fractions is a little trickier, but is just as certain. Let's try the repeating decimal, $0.33\dot{3}3\dot{3}$. First we count the number of repeating digits, and in this case there is just one. We now multiply our decimal by 10^1. If there had been two repeating digits we would use an exponent of two, or 10^2. Multiplying 0.33333 by 10^1 we get 3.33333. Notice that we keep the same number of 3s to the right of the decimal point. We can do this because we always have an "infinity" of 3s to the right. When we subtract our original decimal from this larger decimal, watch what happens:

$$3.33333\dot{3}$$
$$0.33333\dot{3}$$
$$\overline{3.00000}$$

All the repeating digits drop out and we're left with the terminating decimal of 3.0. Now 3.0 is 10 of the original decimals less one of them. Hence, ten times the decimal minus one times the decimal = 3.0. This means that the original decimal = $\frac{3}{9}$ or $\frac{1}{3}$, just what we would expect.

We can perform this operation on any infinite repeating decimal to change it into a fraction. Of course, you might object that something strange is going on when we claim that ten times an infinite, repeating decimal moves all the digits one place to the left (which is the same as saying that it moves the decimal point one position to the right). How do we know this works? This question leads to a strange peculiarity of the decimal system that seems to violate common sense. We can ask the question: What exactly is the number, $0.99999\dot{9}$? Is this number just an infinitesimal amount less than the number 1? In fact, $0.99999\dot{9}$ is exactly 1! If we go through the procedure of changing $0.99999\dot{9}$ into a fraction, we can see it happen. Multiply the decimal by 10 and subtract the original.

$$9.99999\dot{9}$$
$$0.99999\dot{9}$$
$$\overline{9.00000}$$

Now we have ten times the decimal minus one times the decimal = 9 or the original decimal equals $\frac{9}{9} = 1$. Yet, we have relied on the notion that we can multiply an infinite decimal and move the digits toward the left. We can see this is a legitimate step if we resort to limits. Let's show the decimal $0.99999\dot{9}$ as an infinite series or

$$0.99999\dot{9} = \tfrac{9}{10} + \tfrac{9}{100} + \tfrac{9}{1000} + \ldots$$

Now if someone claimed that $0.99999\dot{9}$ was less than 1, then it must be less by a definite amount, say ϵ. But whatever value they picked for ϵ, we could add enough terms of the decimal expansion of $0.99999\dot{9}$ to get closer to 1 than ϵ. Hence, 1 must be the limit of the decimal expansion of $0.99999\dot{9}$.

So far, all this playing around has completely ignored the irrational numbers. What, we may ask, is the decimal expression for an irrational number? It is an infinite, nonrepeating decimal. Therefore, an irrational number such as $\sqrt{2}$ has an infinite nonrepeating decimal expansion. How can we ever represent such a number exactly when neither a fraction nor a decimal can be used? As a matter of fact, we cannot. This very characteristic of irrationals makes them unpleasant objects for some people. They are messy, untidy, and incomplete. Yet, on the other hand, they are exotic, with their decimal expansions wandering all about, never falling into a boring, repeating pattern. In fact, some mathematicians have become obsessed with the decimal expansions of common irrational numbers.

The ratio of the circumference of a circle to its diameter is the value π. It turns out that π is also an irrational number. The first 101 digits of its decimal expansion are given below:

3.14159265358979323846264338327950288419716939937510

58209749445923078164062862089986280348253421170679 . . .

Even though mathematicians know that the pattern will never fall into an infinite repeating pattern, they cannot help but wonder if there are not some other kinds of patterns hidden deep in the expansion. For example, the irrational decimal called the "number number" and formed by consecutive natural numbers is 0.12345678910111213. . . . Since this number is irrational, no repeating pattern develops, yet the method of extending the expansion is obvious: Just write down the next natural number as the next set of digits. What about π? Can we find some pattern that will unlock its secrets? By the fall of 1989 different teams of mathematicians had used supercomputers to calculate over a billion digits of π in their efforts to find something unexpected, and have so far failed.[10] Still, they have not given up and the race continues today to see which computing team can crank out the greatest number of digits.

Considering infinite nonrepeating decimals, we might want to ask how frequently different digits appear in their decimal expansion. In other words, do numbers exist such that, if we could count the proportion of 0s, 1s, 2s, and so on, we would find each of the ten numerals appearing 10% of the time. The answer is yes, and these numbers are called "normal numbers." A number whose decimal expansion contains equal proportions for all digits in every base is called an absolutely normal number. While mathematicians know that the vast majority of all numbers are absolutely normal numbers, they cannot test specific numbers to see if they are normal. Hence, we do not know whether π is normal. It is a strange circumstance that those numbers we encounter in our daily lives are almost always rational numbers and almost none of these are normal, while the overwhelming majority of all real numbers are normal. An example of a normal number which is also rational is the repeating decimal number 0.012345678901234567890012345 . . . , which repeats each numeral once each cycle.

Just how do we calculate irrational numbers? In fact, we have already prepared for this when we learned about converging series. To calculate the decimal expansions of irrational numbers we simply compute successive terms in an infinite series. For example, to calculate $\sqrt{2}$ we can use the following series:

$$\sqrt{2} = 1 + \tfrac{1}{2} - \frac{1}{2\cdot4} + \frac{3}{2\cdot4\cdot6} - \frac{3\cdot5}{2\cdot4\cdot6\cdot8} + \frac{3\cdot5\cdot7}{2\cdot4\cdot6\cdot8\cdot10} - \cdots$$

This series looks messy, yet it is easy to compute, for we can rewrite the series so that each successive term is simply the previous term multiplied by a proper fraction.

$$\sqrt{2} = 1 + \tfrac{1}{2} - \{\tfrac{1}{2}\}\{\tfrac{1}{4}\} + \{\tfrac{1}{2\cdot4}\}\{\tfrac{3}{6}\} - \{\tfrac{3}{2\cdot4\cdot6}\}\{\tfrac{5}{8}\} + \{\tfrac{3\cdot5}{2\cdot4\cdot6\cdot8}\}\{\tfrac{7}{10}\} - \cdots$$

Now we can calculate our decimal expansion of $\sqrt{2}$ and see how each term brings us closer to the correct value to six places of 1.41421. Table 5 gives the first fifteen terms of the expansion. After fifteen terms, the series has reached three places of accuracy, hence the process of approximating $\sqrt{2}$ is slow. Of course, a computer can make many thousands of calculations per second, and the approximation of irrationals with modern technology presents no problems. Mathematicians, in order to get their computations to converge quicker, actually use more complex functions than the simple series offered here.

We have previously presented a series that can be used to approximate π.

$$\pi = 4\cdot\{1 - \tfrac{1}{3} + \tfrac{1}{5} - \tfrac{1}{7} + \tfrac{1}{9} - \ldots\}$$

The first fifteen terms of this series are listed in Table 6. With the same number of terms as the previous series, this series gives us only the

TABLE 5.   Approximating $\sqrt{2}$

Term	Term value	Series sum
1	+1.00000	1.00000
2	+0.50000	1.50000
3	−0.12500	1.37500
4	+0.06250	1.43750
5	−0.03906	1.39844
6	+0.02734	1.42578
7	−0.02051	1.40527
8	+0.01611	1.42138
9	−0.01309	1.40829
10	+0.01091	1.41920
11	−0.00927	1.40993
12	+0.00801	1.41794
13	−0.00701	1.41093
14	+0.00620	1.41713
15	−0.00554	1.41159

TABLE 6.   Approximating π

Term	Term value	Series sum
1	+4.0000	4.0000
2	−1.3333	2.6667
3	+0.8000	3.4667
4	−0.5714	2.8953
5	+0.4444	3.3397
6	−0.3636	2.9761
7	+0.3077	3.2838
8	−0.2667	3.0171
9	+0.2353	3.2524
10	−0.2105	3.0419
11	+0.1905	3.2324
12	−0.1739	3.0585
13	+0.1600	3.2185
14	−0.1481	3.0704
15	+0.1379	3.2083

first digit for π; hence, it converges much more slowly than the series used to approximate $\sqrt{2}$.

Even rational numbers are limiting values for series, for Archimedes discovered the following series for $\frac{1}{3}$:

$$\frac{1}{3} = \frac{1}{4} + \frac{1}{16} + \frac{1}{64} + \frac{1}{256} + \ldots$$

A continued fraction is a special infinite series that consists of an integer plus a fraction, and the fraction's denominator is another integer plus a fraction of the same kind. If the fraction continues without end, it is called a nonterminating continued fraction. A nice nonterminating continued fraction is the following:

$$\frac{(1 + \sqrt{5})}{2} = 1 + \cfrac{1}{1 + \cfrac{1}{1 + \cfrac{1}{1 + \cfrac{1}{1 + \ldots}}}}$$

Of course, the value on the left of the equation is our wonderful Golden Mean. It is known that every irrational number can be represented as a nonterminating continued fraction. Therefore, there is a continued fraction available for estimating the value of every irrational number.

We have now covered much ground in our attempt to understand the irrational numbers. We have the elegant definition by Richard Dedekind that puts irrationals on a sound theoretical footing. We know that they are ordered, just like the rational numbers, and that the real number system is closed for all operations except for division by zero. We can manipulate irrationals just like rationals. As it turns out, we have also accounted for all the points on a line, for the real numbers correspond one-to-one with all the points on the number line. What more could we want to know?

One question we have shied away from concerns how many irrational numbers there are. We know there are an infinite number, but are there "more" or "less" irrationals than rationals? Does it make sense to talk of more or less, when both the rationals and irrationals are infinite in number? We are now prepared to explore this question.

The Story of π

Transcendental Numbers

MORE EQUATIONS

So far we have looked at polynomial equations in the form of $a_0x^n + a_1x^{n-1} + \ldots + a_{n-1}x + a_n = 0$, where the a_0 through a_n are the coefficients to the powers of x. We call such equations standard polynomials. However, there are more ways of setting up equations to solve problems than the polynomial form. In particular, we want to consider three other kinds of equations: exponential, logarithmic, and trigonometric equations. Such equations are basic for solving modern problems from refining oil to putting shuttles into orbit.

Exponential equations comprise unknowns that are part of the exponents of numbers. Such an equation might be $12^x = 144$. Here we need to know to what power we must raise 12 to obtain 144. The answer is simply 2. Thus $12^2 = 144$. But in other exponential equations the answer is not so straightforward. For example, $1.0994^x = 17.32$. What is x? Is x a rational number or an irrational number?

You will probably remember logarithms from college or high school algebra. They are the exponents of numbers. For example, the logarithm of 100 to the base of 10 is 2. This means that 10 must have an exponent of 2 to equal 100. This is shown symbolically as $\log_{10} 100 = 2$. Logarithms with a base of 10 are called *common* logarithms, and we frequently dispense showing the 10. Hence, we have $\log 100 = 2$. One nice feature of logarithms is that we can multiply two numbers together by adding their logarithms because of the identity $A^c \cdot A^d = A^{c+d}$. For example, we have $3^2 \cdot 3^5 = 3^{2+5} = 3^7$. Hence, when we have a difficult or messy multiplication problem, we can frequently change it to an easy addition problem with logarithms. We can also solve certain problems that are expressed in logarithmic form. For example, we might have $\log x = 2$ and want to know what x is. In this case, it is simply 100. Logarithms are a convenient way to approach exponential problems.

Trigonometric equations are based on the relationships between the lengths of the sides of a right triangle and the size of its angles. Figure 41 shows a right triangle with the angle θ (Greek theta) and the three sides marked as hypotenuse, adjacent, and opposite. Numerous trigonometric functions are defined by the relationships between θ and the three sides of the triangle, but we will only consider the sine function here. The sine of an angle is defined as the opposite side divided by the hypotenuse. This is written as $\sin \theta$ = opposite/hypotenuse. Hence, $\sin \theta$, just like $\log 100$, is a number. The $\sin 30° = 0.5$, which says that when θ is $30°$ the hypotenuse is twice as long as the opposite side, and their ratio is opposite/hypotenuse $= 0.5$.

What kind of numbers do we get when we plug different values into exponents, logarithms, and sine functions? In some cases, as we have

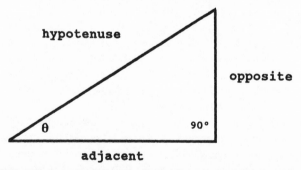

FIGURE 41. Trigonometric relationships involve the various relationships between one angle, θ, in a right triangle and the lengths of the three sides. The sine relationship is defined as $\sin(\theta)$ = (opposite side)/(hypotenuse).

seen above, we get natural numbers or fractions—that is, we get rational numbers. Can we ever get irrational numbers? In fact, most solutions to these kinds of equations are irrational numbers. For example, if θ is a rational number between 0 and 90° then sin θ is irrational except when θ = 30°.

DEFINING THE ELUSIVE *e*

The relationship between a diameter of a circle and its circumference has been known since ancient times and is represented by the symbol π. It was the Swiss mathematician, Leonhard Euler (1707–1783), who actually initiated the use of the Greek letter pi for this relationship (Figure 42). He was also one of the most productive mathematicians who ever lived, publishing over five hundred books and papers and generating enough mathematics to fill ninety volumes.[1] Euler also defined the use of the letter *e* to stand for another fundamental relationship in mathematics:

$$e = \lim_{n \to \infty} \left(1 + \frac{1}{n}\right)^n$$

Hence, *e* is the limit we get when we let *n* grow larger and larger. In other words, we can generate a sequence of numbers based on the formula $(1 + 1/n)^n$ where *n* begins at 1 and increases to each succeeding natural number. This sequence of numbers will get ever closer to the true value of *e*. Hence, *e* is the limit to our sequence and we show this as

$$e = \lim_{n \to \infty} \left(1 + \frac{1}{n}\right)^n$$

Just as we did with π, we can approximate *e* by starting with $n = 1$ and then solving $(1 + 1/n)^n$ for larger *n* (Table 7).

If we continue this process we would arrive at the following value approximated to ten decimal places: $e = 2.7182818284. \ldots$ This number has a beautiful series associated with it that was first discovered by Isaac Newton (1642–1727) in 1665, when he was only twenty-three. A series, you will remember, is a sum of terms, rather than a sequence of

FIGURE 42. Leonhard Euler, 1707–1783. (Photograph from Brown Brothers, Sterling, PA.)

terms. Before we show this wonderful series, we should define a special convention frequently used in mathematics. If we multiply the numbers 1 and 2 we have $1 \cdot 2 = 2$. If we multiply the first three numbers we get: $1 \cdot 2 \cdot 3 = 6$, and multiplying the first four numbers is $1 \cdot 2 \cdot 3 \cdot 4 = 24$. A shorthand for writing these multiplications is the *factorial* sign which is '!'. Hence,

TABLE 7.   Approximating *e*

n	$(1 + 1/n)^n$	Approx. *e*
1	$(1 + \tfrac{1}{1})^1$	2.000
2	$(1 + \tfrac{1}{2})^2$	2.250
3	$(1 + \tfrac{1}{3})^3$	2.370
4	$(1 + \tfrac{1}{4})^4$	2.441
5	$(1 + \tfrac{1}{5})^5$	2.488

$1 \cdot 2 = 2! = 2$ and $1 \cdot 2 \cdot 3 = 3! = 6$, while $1 \cdot 2 \cdot 3 \cdot 4 = 4! = 24$. You can see that the various factorial values grow rapidly. For example $10! = 3{,}628{,}800$. Factorial values are encountered so frequently in mathematics that it is useful to have this shorthand convention for them.

The series that Newton discovered for *e* is the following:

$$e = \frac{1}{0!} + \frac{1}{1!} + \frac{1}{2!} + \frac{1}{3!} + \frac{1}{4!} + \ldots$$

The symbol $0!$ is defined to be 1. The above series can be written as: $e = 1 + 1 + \tfrac{1}{2} + \tfrac{1}{6} + \tfrac{1}{24} + \ldots$ and can be represented as $e = \Sigma(1/n!)$ where *n* assumes the values of whole numbers beginning with zero. You will remember that the Greek capital letter sigma (Σ) means we add up all the terms in our series.

The reason mathematicians and scientists are interested in *e* is that it crops up in so many different places when trying to solve problems. This is related to the beautiful property that for small values of *x* we have $e^x \approx 1 + x$. Here we have used the special approximation symbol $\approx$ to show that the two sides of the equation are close but not necessarily equal. We can see that the above relationship holds if we let *x* be equal to $1/n$ where *n* is large ($e^{1/n} \approx 1 + 1/n$). Now we raise each side to the *n*th power to obtain $(e^{1/n})^n \approx (1 + 1/n)^n$ or $e \approx (1 + 1/n)^n$, which is just how we defined the limit as *n* increases to infinity.

What in the world does *e* have to do with our search for numbers? When mathematicians began asking just what kinds of numbers *e* and π were, they discovered a completely new class of numbers.

LOOKING AT π AND e

If we have a polynomial equation in the first degree or $ax + b = 0$, where a and b are natural numbers, then the solutions for x are always rational numbers since $x = -b/a$. In fact, all rational numbers are roots of some first degree polynomial equation. We have seen that polynomials of higher order can have both rational and irrational solutions. We define the solutions to polynomials where the coefficients are integers as *algebraic* numbers. To be precise, we can say that all the solutions to polynomials with rational coefficients are algebraic. This is because for any polynomial equation with fractions (rational numbers) as coefficients we can find an integer to multiply through both sides of the equation and convert it to a polynomial with integer coefficients. For example, $\frac{1}{2}x + \frac{2}{3} = 0$ can be changed to a polynomial with whole number coefficients by multiplying both sides of the equation by 6 or: $6 \cdot (\frac{1}{2}x + \frac{2}{3}) = 6 \cdot 0$ becomes $6 \cdot (\frac{1}{2}x) + 6 \cdot (\frac{2}{3}) = 0$ or $3x + 4 = 0$.

Both the equations $\frac{1}{2}x + \frac{2}{3} = 0$ and $3x + 4 = 0$ have the same solution, which is $x = -\frac{4}{3}$. Therefore, we can take any polynomial equation with rational coefficients (fractions) and change it to a polynomial equation with whole number or integer coefficients, and the solutions stay the same. This means that the natural numbers, fractions, and radicals $(\sqrt{n})$ are all algebraic numbers, since they are the solutions to such polynomials.

> DEFINITION: An *algebraic* number is a number that is a solution
> to a polynomial equation whose coefficients are all integers.

It was Euler who first asked in 1748 whether e and π are algebraic numbers. That is, can they be the solutions of polynomials with integer coefficients? This question had a special significance for π. Since the time of the ancient Greeks, an unanswered question asked whether it is possible to construct a square with the same area as a given circle using only a straight edge and a compass. The area of a circle is given by the formula πr^2 where r is the radius of the circle. If we begin with a circle with radius $= 1$ then the area is just π. To construct a square with an area of π we must construct a line equal in length to $\sqrt{\pi}$ so that when it is squared we get the desired area.

This problem is called "squaring the circle" and has been attempted by countless amateur and professional mathematicians for the last two thousand years. If we limit ourselves to a straight edge and compass, then the various ways we can manipulate the lengths associated with a circle are restricted to multiplication, addition, subtraction, and division. These are the four operations used in our standard polynomial. Therefore, if π is a solution to a standard polynomial, then it should be possible to square the circle, and π would be an algebraic number. In other words, if π is algebraic, then the circle can be squared. Otherwise, it cannot.

If e and π are not algebraic, then what could they be? Mathematicians began calling nonalgebraic numbers *transcendental* numbers. Yet, when Euler first raised the question, it was not known whether such numbers even existed. Although this question was first asked in 1748, it was not answered until 1844, long after Euler's death. In that year the French mathematician Joseph Liouville (1809–1882) constructed the first number proven to be transcendental. His number was

$$L = \frac{1}{10^{1!}} + \frac{1}{10^{2!}} + \frac{1}{10^{3!}} + \frac{1}{10^{4!}} + \ldots \text{ or}$$

$$L = \frac{1}{10} + \frac{1}{10^2} + \frac{1}{10^6} + \frac{1}{10^{24}} + \ldots$$

In decimal form we get $L = 0.1100010000000000000000000100. \ldots$ Liouville was able to prove that this number could not be the solution to any polynomial equation with integer coefficients and was therefore one of the elusive transcendental numbers. He then went on to show how to construct an infinite number of transcendental numbers of the following form

$$\frac{a_1}{10^{1!}} + \frac{a_2}{10^{2!}} + \frac{a_3}{10^{3!}} + \ldots$$

where the different a's are integers in the range of 0 to 9. Such numbers are called Liouville numbers. This settled once and for all the question of whether transcendental numbers really existed. But still no one knew the status of numbers such as π and e; were they transcendental or algebraic? Finally, in 1873, Charles Hermite (1822–1901) proved that e was transcendental. In 1882, Ferdinand von Lindemann (1852–1939) proved that

π was transcendental. This settled forever the question of squaring the circle, proving that it was an impossibility.

Given that π and e and the Liouville numbers are transcendental, what about other possible candidates? The exponential and logarithmic equations generally have solutions which are transcendental. However, trying to determine if a particular solution is transcendental or algebraic can be a daunting task. Certain other combinations are still unknown. While it has been proven that e^π is transcendental, it is not known whether e^e or π^π or π^e are transcendental or algebraic. In addition, the status of such simple expressions as $e + \pi$ and $e \cdot \pi$ is still unknown.

We now know there exists a whole class of exponential numbers that are transcendental. In 1934 the Russian mathematician Aleksander (Alexis) Osipovich Gelfond (1906–68) proved that all the numbers with the form a^b are transcendental if a is algebraic and not 0 or 1, and b is an irrational algebraic number. This is now called Gelfond's theorem (or sometimes the Gelfond–Schneider theorem since Theodor Schneider independently proved the theorem in 1935), and means that numbers such as $3^{\sqrt{7}}$ and $(\sqrt{6})^{\sqrt{5}}$ are transcendental numbers since the bases (3 and $\sqrt{6}$) are algebraic and not 0 or 1, and the exponents ($\sqrt{7}$ and $\sqrt{5}$) are irrational, algebraic numbers. Unfortunately, this does not help us decide the status of numbers such as e^e, π^π, or π^e since here both the bases and exponents are already transcendental numbers and do not satisfy Gelfond's conditions.

Also from Gelfond's theorem we can deduce that a number such as $\log 2$ must be transcendental. Log 2 is the power to which 10 must be raised to obtain 2. Hence, we have $10^{\log 2} = 2$. We can prove that $\log 2$ is an irrational number. Let's suppose that $\log 2$ is rational. This would mean there exist two integers p and q such that $\log 2 = p/q$. From the definition of logarithms we get $10^{\log 2} = 10^{p/q} = 2$. Now we raise both sides of the equation to the q power and get $10^p = 2^q$. Now 10 is the product of 2 times 5 or $(2 \cdot 5)^p = 2^q$ or $2^p \cdot 5^p = 2^q$.

Now we have two possibilities. If p is larger than q, then we can cancel 2^q out of both sides of the equations to get $2^{p-q} \cdot 5^q = 1$, which is clearly false. What if q is larger then p? In this case we cancel 2^p out of both sides of the equation to get $5^p = 2^{q-p}$. This also must be false since all composite numbers have a unique factorization. In other words, no number exists that can be written both as 5 raised to an integral power and as 2 raised to an integral power. We realize this at once when we remember that 2 raised to any whole number power will be an even number, and 5 raised to any whole number power will be an odd number (will always be a number whose last digit is 5). Hence, we cannot represent a power of 5 as a power of 2. Therefore, p and q do not exist and $\log 2$ is irrational.

Now, if $\log 2$ were both irrational and algebraic then we would know from Gelfond's theorem that 2 must be transcendental because $10^{\log 2} = 2$. But 2 is rational, so $\log 2$ must be both irrational and transcendental. This same argument works for any $\log x$ where x is rational and $\log x$ is irrational. These conditions exist for the overwhelming majority of values for x. Hence, almost all logarithms are transcendental. On the other hand, while standard trigonometric functions are generally irrational, they are also algebraic and not transcendental.

HOW MANY TRANSCENDENTAL NUMBERS ARE THERE?

We know that many transcendental numbers exist. Certainly, there are an infinite number of them. Are there more transcendental numbers than algebraic numbers? Does it even make sense to talk of one infinite collection of numbers being larger or smaller than another? The struggle to answer these questions opened an entirely new vista of mathematical thought regarding the infinite, and provided us with a startling insight to the nature of the infinite.

The first infinite collection of things human beings came in contact with was the set of natural numbers. If we are going to get a grip on infinity, then understanding the natural numbers is the best place to start. When two finite sets have the same cardinal number, we said that one set could be mapped one-to-one onto the other set. If no elements are left over in either set, then they have the same cardinal number. We will use the same principle when discussing infinite sets, and use mapping as a way to compare them.

DEFINITION: If A and B are two infinite sets, and if all the elements of A map one-to-one onto all the elements of B, then A and B have the same cardinal number.

Since the two sets A and B are infinite, we cannot use any finite number as their cardinal number. Hence, we will eventually find it necessary to define some additional symbols to represent cardinal numbers of infinite sets.

Any infinite set that can be put into a one-to-one mapping with the natural numbers we call a countable set. As it turns out, countable sets have some remarkable properties. Galileo Galilei (1564–1642) is one of the most famous scientists and mathematicians. He improved the telescope and discovered the satellites of Jupiter, the rotation of the sun, and sunspots. He discovered the law of the pendulum and proved the uniform acceleration of bodies falling to earth. In Galileo's *Dialogues Concerning Two New Sciences*, published in 1636, he pointed out a very curious feature of infinite sets.[2] He demonstrated that it is possible to make a one-to-one mapping of the natural numbers onto the squares of natural numbers. Such a mapping is:

1	2	3	4	5	6	7	8	9	10	11	...
↕	↕	↕	↕	↕	↕	↕	↕	↕	↕	↕	
1	4	9	16	25	36	49	64	81	100	121	...

Each natural number is mapped to only one square number, and each square number is mapped to only one natural number. Thus, we have a one-to-one mapping. Every natural number and every square number is accounted for. From this, we get the startling conclusion that there are just as many square numbers as there are natural numbers! The natural numbers are called countable (or enumerable), and any other set that can be mapped one-to-one with them is also called countable.

When we first see this, we are tempted to throw up our hands and declare there must be a mistake. How could there be as many squares as natural numbers when the infinite set of squares is missing so many natural numbers? We intuitively feel that the natural numbers must be "more numerous" than the square numbers. But according to the definition given above, the two sets of numbers have the same cardinal number and are, thus, equal in "size." This little illustration points out a remarkable characteristic about countable, infinite sets. We can make the same argument regarding all positive and negative integers and the natural numbers with the following mapping:

1	2	3	4	5	6	7	8	9	10	11	...
↕	↕	↕	↕	↕	↕	↕	↕	↕	↕	↕	
0	1	−1	2	−2	3	−3	4	−4	5	−5	...

Here, again, we have achieved a one-to-one mapping and accounted for every integer and every natural number. Hence, there are just as many natural numbers as integers. In fact, we have the following nice theorem.

THEOREM: Every countable infinite set has the same cardinal number as the natural numbers.

Our theorem simply says that when we talk of the size of countable infinite sets, all such sets are the same size, that is, they have the same cardinal number. This theorem implies that an infinite subset of a countable set is the same size as the original set. One reason this idea gives us a funny feeling is a property of finite numbers that infinite numbers do not share. For example, if we add 1 to 10 we get a number larger than 10. But, if we add 1 to an infinite set, we do not get a "bigger" set, we get a set of the same size. Hence, if A is the cardinal number for an infinite set then $A + 1 = A$. This looks entirely wrong, for in our everyday experience when we add an object to a collection, the collection grows larger. But we must realize that we are dealing with infinite sets, and they do not always behave the same way as finite sets. One way to reconcile this behavior is to think of an infinite set as so large that adding 1 (or any finite number) to it is too insignificant to change its size. To really change the size of an infinite set, we must make some infinite change. If you are confused by this idea that adding a finite amount to an infinite set does not change its size, then you are not alone. Mathematicians have been puzzling over this for decades. When we talk about the size of an infinite set, remember that we are defining size in terms of mappings. By using the notion of mapping we can see that infinite sets really do come in different sizes, but these differences are not finite, they are infinite differences.

Now that we have our procedure for mapping we can proceed to one of the most astounding conclusions of mathematics. We know that any infinite set mapped one-to-one with the natural numbers has the same cardinal number as the set of natural numbers. How many rational numbers are there? Can we map all the rational numbers onto the natural numbers? At first glance, it would seem that the set of all rational numbers must be much larger than the natural numbers. Between the natural numbers 1 and 2 alone there exists an infinity of rational numbers. We can generalize this and say that between every two natural numbers there exists an infinity of rational numbers. How then could the two sets have the same cardinal number?

Consider the mapping in Table 8 where we place every rational number in a square matrix. This matrix continues infinitely to the right and infinitely down, and accounts for every rational number, since every rational number is of the form p/q where p and q are whole numbers. We pass through the fractions beginning in the upper left corner and eliminate duplicates to get the matrix in Table 9. In this matrix the arrows show how we are to proceed through the numbers, eliminating duplicates. The decision to eliminate a fraction is not difficult. If the fraction is already reduced to its lowest terms, then we keep it, if not, we eliminate it. Every positive rational number will be represented in the matrix. If we want to find p/q (assuming this ratio is in its lowest terms), then we simply go down to row p and out to column q. Hence, all positive rational numbers

TABLE 8.   Matrix of Rational Numbers

$\frac{1}{1}$	$\frac{1}{2}$	$\frac{1}{3}$	$\frac{1}{4}$	$\frac{1}{5}$	$\frac{1}{6}$	$\frac{1}{7}$	$\frac{1}{8}$	$\cdots$
$\frac{2}{1}$	$\frac{2}{2}$	$\frac{2}{3}$	$\frac{2}{4}$	$\frac{2}{5}$	$\frac{2}{6}$	$\frac{2}{7}$	$\frac{2}{8}$	$\cdots$
$\frac{3}{1}$	$\frac{3}{2}$	$\frac{3}{3}$	$\frac{3}{4}$	$\frac{3}{5}$	$\frac{3}{6}$	$\frac{3}{7}$	$\frac{3}{8}$	$\cdots$
$\frac{4}{1}$	$\frac{4}{2}$	$\frac{4}{3}$	$\frac{4}{4}$	$\frac{4}{5}$	$\frac{4}{6}$	$\frac{4}{7}$	$\frac{4}{8}$	$\cdots$
$\frac{5}{1}$	$\frac{5}{2}$	$\frac{5}{3}$	$\frac{5}{4}$	$\frac{5}{5}$	$\frac{5}{6}$	$\frac{5}{7}$	$\frac{5}{8}$	$\cdots$
$\frac{6}{1}$	$\frac{6}{2}$	$\frac{6}{3}$	$\frac{6}{4}$	$\frac{6}{5}$	$\frac{6}{6}$	$\frac{6}{7}$	$\frac{6}{8}$	$\cdots$
$\frac{7}{1}$	$\frac{7}{2}$	$\frac{7}{3}$	$\frac{7}{4}$	$\frac{7}{5}$	$\frac{7}{6}$	$\frac{7}{7}$	$\frac{7}{8}$	$\cdots$

.
.
.

TABLE 9. Matrix of Rational Numbers

$$\frac{1}{1} \rightarrow \frac{1}{2} \quad \frac{1}{3} \rightarrow \frac{1}{4} \quad \frac{1}{5} \rightarrow \frac{1}{6} \quad \frac{1}{7} \rightarrow \frac{1}{8} \quad \cdots$$

$$\frac{2}{1} \quad \frac{2}{3} \quad \frac{2}{5} \quad \frac{2}{7} \quad \cdots$$

$$\frac{3}{1} \quad \frac{3}{2} \quad \frac{3}{4} \quad \frac{3}{5} \quad \frac{3}{7} \quad \frac{3}{8} \quad \cdots$$

$$\frac{4}{1} \quad \frac{4}{3} \quad \frac{4}{5} \quad \frac{4}{7} \quad \cdots$$

$$\frac{5}{1} \quad \frac{5}{2} \quad \frac{5}{3} \quad \frac{5}{4} \quad \frac{5}{6} \quad \frac{5}{7} \quad \frac{5}{8} \quad \cdots$$

$$\frac{6}{1} \quad \frac{6}{5} \quad \frac{6}{7} \quad \cdots$$

$$\frac{7}{1} \quad \frac{7}{2} \quad \frac{7}{3} \quad \frac{7}{4} \quad \frac{7}{5} \quad \frac{7}{6} \quad \frac{7}{8} \quad \cdots$$

are included. We can now map these numbers one-to-one with the natural numbers by taking them in order.

$$\frac{1}{1} \quad \frac{1}{2} \quad \frac{2}{1} \quad \frac{3}{1} \quad \frac{1}{3} \quad \frac{1}{4} \quad \frac{2}{3} \quad \frac{3}{2} \quad \frac{4}{1} \quad \frac{5}{1} \quad \cdots$$
$$\updownarrow \quad \updownarrow \quad \updownarrow \quad \updownarrow \quad \updownarrow \quad \updownarrow \quad \updownarrow \quad \updownarrow \quad \updownarrow \quad \updownarrow$$
$$1 \quad 2 \quad 3 \quad 4 \quad 5 \quad 6 \quad 7 \quad 8 \quad 9 \quad 10 \quad \cdots$$

You might object here that we have only accounted for positive rational numbers. That is true, but we can also construct the same matrix with negative fractions. We then include zero as the first element and take alternate fractions from the positive and negative matrices to get

$$0 \quad \frac{1}{1} \quad -\frac{1}{1} \quad \frac{1}{2} \quad -\frac{1}{2} \quad \frac{2}{1} \quad -\frac{2}{1} \quad \frac{3}{1} \quad -\frac{3}{1} \quad \frac{1}{3} \quad \cdots$$
$$\updownarrow \quad \updownarrow \quad \updownarrow \quad \updownarrow \quad \updownarrow \quad \updownarrow \quad \updownarrow \quad \updownarrow \quad \updownarrow \quad \updownarrow$$
$$1 \quad 2 \quad 3 \quad 4 \quad 5 \quad 6 \quad 7 \quad 8 \quad 9 \quad 10 \quad \cdots$$

Now we have the mapping we want. Every rational number, positive, negative, and zero, is included only once and these are mapped onto the natural numbers. Therefore, we have the startling conclusion that the cardinal number for the natural numbers is equal to that for all rational numbers. What about the set of all algebraic numbers? Or all transcendental numbers? Maybe there is only one cardinal number for all infinite sets!

Since Galileo first pointed out the mapping between natural numbers and square numbers, mathematicians have puzzled over what the implications of this are. The full answer to this question had to wait for approximately 250 years, or until the end of the nineteenth century.

THE REMARKABLE GEORG CANTOR

The balance of what we have to learn about transcendental numbers is almost entirely due to one remarkable man, Georg Cantor, who led a fascinating but tragic life (Figure 43). He was born in St. Petersburg, Russia, on March 3, 1845.[3] His father was Georg Woldeman Cantor, a Danish merchant, and his mother was Maria Böhm Cantor. In 1856, when Georg was eleven, the family resettled in Frankfurt am Main in Hesse so that Georg Senior could avoid the harsh Russian winters. When young Georg was fifteen he demonstrated a strong talent for mathematics and was enrolled in the Grand-Ducal Higher Polytechnic in Darmstadt where he was to study, at his father's request, engineering. In 1863, he entered the University of Berlin, where he chose to study mathematics, physics, and philosophy. He was a devout young man, a characteristic he inherited from his parents.

Teaching at the mathematics department at the University of Berlin were two of the greatest nineteenth century mathematicians: Karl Weierstrass (1815–1897) and Leopold Kronecker (1823–1891). These two famous men would play decisive roles in Cantor's life, one for good, the second for ill. While Weierstrass would become one of Cantor's supporters, Kronecker would carry on a lifelong battle against Cantor's ideas.

Upon graduating *magna cum laude* with his doctorate from the University of Berlin in 1867, Cantor could not find a university position commensurate with his talents. Therefore, he accepted a job teaching mathematics to young women at a private school. In 1869, Cantor found a faculty opening at the small institution of Halle University, still not a position which matched his fine training at the University of Berlin.

FIGURE 43. Georg Cantor, 1845–1918. (Photograph from Brown Brothers, Sterling, PA.)

However, he settled in at Halle University and was made an assistant professor in 1872. During that same year Cantor met and befriended another young German mathematician. His name was Richard Dedekind— the very man who gave us our modern definition of irrational numbers, publishing his famous paper the year they met! These two mathematicians, who would add so much to mathematics at the end of the nineteenth century, became close friends, sharing and supporting each other's ideas.

ARE THE ALGEBRAIC NUMBERS COUNTABLE?

The algebraic numbers include all the rational numbers plus those irrational numbers that are the solutions to polynomials. Is the set of algebraic numbers countable? This would seem a much more difficult question, but Cantor solved it easily. What he did was develop an ordering of algebraic numbers by considering the polynomials they are the solutions for.

Cantor defined an integer associated with each polynomial, which he called the polynomial's height, H. Remember that a polynomial has the form $a_0 x^n + a_1 x^{n-1} + a_2 x^{n-2} + \ldots + a_{n-1} x + a_n = 0$. We define its height in the following way: First we compute $n-1$ which is just the largest exponent of x minus 1. Then we add to this the absolute value of all the coefficients. By absolute value we mean the positive value of the coefficient whether the value is positive or negative in the polynomial. This makes the polynomial's height equal to $H = (n - 1) + |a_0| + |a_1| + \ldots + |a_n|$, where the bars bracketing the coefficients call for their absolute values. For example, the simple polynomial, $x + 1 = 0$ has a height of 2, because $a_0 = 1$, $a_1 = 1$, and $n = 1$. This gives us $H = (1 - 1) + 1 + 1 = 2$. We can see at once that for each value of H we have only a finite number of possible polynomials. Table 10 gives the associated polynomials with the first three values for H.

For $H = 1$, we have only one polynomial, $x = 0$. When H is 2, we have four associated polynomials, and when it is 3 we get eleven different polynomials. Hence, the number of polynomials associated with each height increases rapidly. Yet, only a finite number of polynomials is associated with any particular H. Now we use the famous theorem by Carl Gauss, the Fundamental Theorem of Algebra. It says that every polynomial equation has at least one solution or root. A corollary to his theorem is that a polynomial has exactly the same number of solutions as the degree of the polynomial. Hence, each polynomial has a fixed number of solutions; for example, the polynomial $x^3 + 2x - 5 = 0$ has three solutions while the polynomial $x^2 - 7x + 1 = 0$ has two solutions. For

TABLE 10.   Polynomial Heights

Height	Polynomials	Unique solutions
1	$x = 0$	0
2	$x + 1 = 0$, $x - 1 = 0$, $2x = 0$, $x^2 = 0$	-1, 1
3	$x + 2 = 0$, $x - 2 = 0$, $2x + 1 = 0$, $2x - 1 = 0$, $3x = 0$, $x^2 + 1 = 0$, $x^2 - 1 = 0$, $x^2 + x = 0$, $x^2 - x = 0$, $2x^2 = 0$, $x^3 = 0$	-2, 2, $-\frac{1}{2}$, $\frac{1}{2}$, $\sqrt{-1}$, $-\sqrt{-1}$

each polynomial in Table 10 we can write down the solutions and eliminate the duplicates. Then, for each set of polynomials associated with each value of H, we have a fixed number of unique solutions, or algebraic numbers.

From Table 10 we see that there is only one unique solution for $H = 1$, that is, zero. For $H = 2$ we get two solutions, 1 and -1. For $H = 3$ we get six unique solutions. For the moment we are going to ignore the fact that we do not have a clue what kind of numbers $\sqrt{-1}$ and $-\sqrt{-1}$ are (we will learn all about them later). We can now map our unique solutions onto the natural numbers in the following manner.

$$
\begin{array}{ccccccccc}
0 & -1 & 1 & -2 & 2 & -\frac{1}{2} & \frac{1}{2} & \sqrt{-1} & -\sqrt{-1} & \cdots \\
\updownarrow & \updownarrow & \updownarrow & \updownarrow & \updownarrow & \updownarrow & \updownarrow & \updownarrow & \updownarrow & \\
1 & 2 & 3 & 4 & 5 & 6 & 7 & 8 & 9 & \cdots
\end{array}
$$

In the above mapping every algebraic number is accounted for, because every polynomial has an associated H value. Associated with that H value will be the unique algebraic numbers to be mapped onto the natural numbers. From this one-to-one mapping we realize, just as Cantor did, that the algebraic numbers must be countable. Are we ever to find an infinite set that is not countable?

HOW MANY TRANSCENDENTAL NUMBERS?

Cantor was now ready to pose the most important question of his life: Are the transcendental numbers countable? He asked this question asking whether the real numbers, which include the algebraic and transcendental numbers, are countable. If the transcendental and the algebraic numbers are countable, then their sum would also be countable. On November 29, 1873, Cantor wrote a letter to his friend, Richard Dedekind.

> May I ask you a question, which has a certain theoretical interest for me, but which I cannot answer; maybe you can answer it and would be so kind as to write to me about it. It goes as follows: take the set of all natural numbers n and denote it N. Further, consider, say, the set of all positive real numbers x and denote it R. Then the question is simply this: can N be paired with R in such a way that to every individual of one set corresponds one and only one individual of the other? At first glance, one says to oneself, "No, this is impossible, for N consists of discrete parts and R is a continuum." But nothing is proved by this objection. And much as I too feel that N and R do not permit such a pairing, I still cannot find the reason. And it is this reason that bothers me; maybe it is very simple.[4]

Here, Cantor has ignored the negative real numbers, but this is not a problem. If he can find a mapping of all positive real numbers onto the natural numbers, then, by extension of the same technique, he can account for all negative real numbers too. Dedekind wrote back to Cantor that he did not know the answer to Cantor's question. Then, on December 7, 1873, just eight days after the first letter, Cantor wrote again to Dedekind.

> Recently I had time to follow up a little more fully the conjecture which I mentioned to you; only today I believe I have finished the matter. Should I have been deceived, I would not find a more lenient judge than you. I thus take the liberty of submitting to your judgment what I have written, in all the incompleteness of a first draft.[5]

In the space of little more than a week, Cantor had stumbled across the discovery of his lifetime—that the real numbers are not countable because there are too many of them. We will actually look at two of his proofs. The first is the proof he offered to Richard Dedekind in December of 1873 and published in 1874.[6] The second proof is a clever application of the decimal system which is found in most texts on set theory. Both proofs are beautiful, but the first is somewhat more elegant.

In his first proof, Cantor used the method introduced by the Greeks, *reductio ad absurdum*, or the indirect method of proof, which we have previously encountered. Basically, he assumed that the positive real numbers could be mapped one-to-one onto the natural numbers and showed this led to a contradiction. If the set of positive real numbers is uncountable, then the set of all positive and negative real numbers is also uncountable. This proof also suffices to prove that the transcendental numbers are uncountable, since all the other kinds of real numbers are countable. If the set of all real numbers is uncountable, it must be due to the transcendental numbers.

If we assume that the positive real numbers are countable, then there must exist at least one mapping between them and the natural numbers. Assume that such a mapping is given by the following sequence where each ω (omega) is a positive real number:

$$\omega_1, \ \omega_2, \ \omega_3, \ \omega_4, \ \omega_5, \ \omega_6, \ \omega_7, \ \omega_8, \ \omega_9, \ \ldots$$

In the above sequence, the subscripts indicate the natural numbers that the real numbers, the various ω, are mapped onto. From Cantor's assumption, the above sequence contains all positive real numbers.

Let's take a segment on the real number line, and we will designate the end points as α and β. It really does not matter which two numbers we choose as the segment's end points. All we assume here is that α and β are different numbers, and we will arbitrarily say that α is the smaller of the two so that $\alpha < \beta$. Since all real numbers are contained in our sequence, then we know that all the real numbers between α and β must also be in the above sequence. Let's move along the sequence until we encounter two numbers, we will call them ω_a and ω_b, that fall inside our segment, (α, β). We will also say for the sake of argument that $\omega_a < \omega_b$. These two numbers, selected from our sequence, define another line segment inside our first segment (Figure 44). Now we continue along our sequence until we find two more numbers, but this time they are between ω_a and ω_b and we will call these new numbers ω_c and ω_d. These two new numbers define a third segment inside the segment (ω_a, ω_b). As you can see, we can continue this procedure indefinitely. Each time we define a new line segment by finding two more numbers in our sequence, we can, in turn, use this line segment to define another segment. The first few line segments have been illustrated in Figure 44.

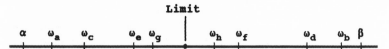

FIGURE 44. Cantor's first proof of the uncountability of real numbers. If ω_1, ω_2, ω_3, . . . is a countable sequence of real numbers, then it is possible to construct an infinite set of nested intervals ($\{\omega_a, \omega_b\}$, $\{\omega_c, \omega_d\}$, $\{\omega_e, \omega_f\}$) that has a limit point that is not in the sequence. Hence, no countable sequnce can contain all real numbers.

Now two things will happen as this process continues. First, we might assume that the process stops, and we never find another two numbers that fit in our last defined line segment. But this is a contradiction, because we know that between any two real numbers there is another number. In fact, we can go further than that: between any two real numbers there are an infinity of real numbers. Therefore, this first case cannot happen. No matter how many times we have found smaller and smaller line segments, all fitting inside each other, we can continue in our real number sequence and find another two numbers that fall inside our last segment.

Therefore, the second condition must exist: We can define an infinite number of segments, each inside the other. Similarly, these segments will have an infinite number of end points (numbers) all confined inside the original segment (α,β). At this point in his proof Cantor used a theorem that was discovered earlier in the nineteenth century by two mathematicians: the Czechoslovakian priest, Bernhard Bolzano (1781–1848), and one of Cantor's teachers at the University of Berlin, Karl Weierstrass. It was Weierstrass who presented the proof of the Bolzano–Weierstrass Theorem in 1865, long after the death of Bolzano, yet both are given credit for the essential ideas involved. The Bolzano–Weierstrass Theorem that Cantor needed is the following:

> BOLZANO–WEIERSTRASS THEOREM: If x_1, x_2, x_3, . . . is an increasing sequence of numbers that is bounded by some number B larger than every x, then there exists a limit, L, to the sequence.

Now when we look at the infinite collection of segments we have defined from our sequence of positive real numbers, we see at once that the lower end points of our segments form a new sequence of numbers, an infinite, increasing sequence bounded by β above. Hence, between α and β there exists a limit point, L, to this sequence. The sequence is simply: $\alpha, \omega_a, \omega_c, \omega_e, \ldots$ that we discovered in the original sequence of ω. What kind of number is L? It cannot be one of our real numbers in the sequence $\omega_1, \omega_2, \omega_3, \omega_4, \ldots$ because, if it were, we would eventually come to it in our search for numbers to define additional segments, and L would become the end point of a segment. This notion takes a moment of reflection to appreciate. If L were in the sequence of real numbers then L would be mapped onto a specific natural number, n, and have this number as its subscript. If our search for more segments inside our original segment continued forever, then in n steps we would come to L_n and L would then become one of our segment end points, and therefore not our limit.

Therefore, since L cannot be in the sequence of ω and at the same time be a limit, we have found a real number (since L is a number on our number line) that was not included in the sequence ω. Therefore, the sequence ω does not contain all positive real numbers. This contradiction proves that no sequence of real numbers mapped one-to-one with the natural numbers can be complete, and countless real numbers will be excluded. Hence, the cardinal number of real numbers is "larger" than the cardinal number for the natural numbers (or the rational or even the algebraic numbers).

A SECOND PROOF

The first proof given by Cantor in 1873 that the positive real numbers are not countable has a certain elegance since it does not rely on anything but the notion of mapping numbers and the Bolzano–Weierstrass Theorem. The second proof, developed some years later, relies on our decimal system. This proof is essentially the same as the first proof since it uses the indirect method where we assume that a one-to-one mapping is possible between the real numbers and the natural numbers. In this case we are going to assume that our numbers are represented by their decimal forms. In addition, we will assume that we have mapped only the real numbers between zero and 1. If we can prove that the real numbers between zero and 1 are uncountable, then certainly all the real numbers are uncountable.

Suppose we have listed all real numbers between zero and 1 in the following mapping:

$$1 \leftrightarrow 0.1792038409827 \ldots$$
$$2 \leftrightarrow 0.3755500000000 \ldots$$
$$3 \leftrightarrow 0.0001788345441 \ldots$$
$$4 \leftrightarrow 0.4998783333333 \ldots$$
$$5 \leftrightarrow 0.8455967928739 \ldots$$
$$\vdots$$

Now, of course, the list goes on forever, since there are infinitely many natural numbers and infinitely many real numbers between zero and 1. Some of the real numbers in the list will be rational numbers with terminating or repeating decimals while others will be irrational numbers that are nonrepeating. The assumption is, of course, that all real numbers between zero and 1 will be on this list.

To prove a contradiction, all we have to do is construct a number between zero and 1 that is not on the list. This is how we do it. In our first decimal place to the right of the decimal point we pick a digit different from the one shown for the number mapped to 1. That is, we pick a digit different than 1. Let's pick the number 3, which means our constructed number begins with 0.3, . . . For our second digit we go to the second real number on the list and choose a digit different from the second digit of the second number. Hence, we choose a digit different from 7. Let's make it 4. Now our constructed number is 0.34. . . . We continue in this fashion, always selecting the nth digit to be different from the nth digit of the nth real number. When this number is constructed we can see at once that it cannot be a number in our list. Why? Suppose someone claimed that the nth number on the list was our constructed number. We could go to the nth digit in that nth number and point out that it was different from our nth digit. Hence, the list of real numbers cannot be complete, just as the sequence of ω was not complete from the previous proof. That is, there are too many real numbers to map them onto the natural numbers. We have the following beautiful theorem, which we credit to Cantor. Why is it beautiful? It is so short, yet so powerful.

THEOREM: The set of all real numbers is not countable.

Cantor gave names to the cardinal numbers for both the set of natural numbers and the set of real numbers. The natural numbers (and by extension all countable sets) has the cardinal number $\aleph_0$ which is simply the first letter of the Hebrew alphabet, aleph, with a subscript of zero. The cardinal number for the real numbers he named $\aleph_1$. Thanks to Cantor we have now discovered a deep truth about numbers. We know that the sets of all natural numbers, all fractions, and all algebraic numbers have the same cardinal number, $\aleph_0$, but that the other numbers on the number line, the transcendental numbers, have a greater density, and they form a set of $\aleph_1$ elements.

Here we can take a moment to ponder just what is suggested when we say the transcendental numbers are uncountable. Consider for a moment that we gave a specific definition to the transcendental number, e. We said that $e = \lim (1 + 1/n)^n$. Hence, we described e with a finite number of symbols. Can we define or describe each and every transcendental number with a finite set of symbols? The answer is no. If we could, then we could arrange the set of all finite descriptions into a countable set. Hence, we could use the finite descriptions as a kind of mapping of the transcendental numbers onto the natural numbers and they would not be uncountable. This leads to the strange conclusion that there must be an uncountable number of transcendental numbers that we can never describe with a finite number of symbols. Hence, there exist an uncountable number of numbers that we can never describe. Does this mean there exist individual numbers that are indescribable? Not necessarily. Specific numbers can always be studied and, we hope, described with a finite set of symbols and added to our countable list of described transcendental numbers.

IS THERE SOMETHING BETWEEN $\aleph_0$ AND $\aleph_1$?

As with much in mathematics, as soon as we answer one question, more pop up to demand answers. Once Cantor proved that there are two distinct infinite sets, $\aleph_0$ and $\aleph_1$, the next question was whether there is any infinite set between them. That is, does there exist an infinite set that is too large to map onto the natural numbers and too small to map onto the real numbers? This question became known as the *continuum hypothesis*. Cantor did not believe such a set existed but could not prove it. The answer to this question had to wait until 1963. In this year the young mathematician Paul Cohen (1934–   ) proved that the question was *undecidable*. Just what does it mean for a question to be undecidable? Normally, we think of "yes" and "no" questions having only two answers, either "yes" or "no." But in questions dealing with mathematics, this is not necessarily so. Logicians like to construct the various fields of mathematics from sets of axioms that they accept as true. This is just what Euclid did when he developed his deductive geometry. He made some claims, which he did not prove, and simply accepted them as intuitively true. Then he went on to use them to deduce theorems of geometry.

How does this relate to undecidability? Euclid's fifth postulate says that for any line and a point not on that line, one and only one line passes through that point that is parallel to the original line (Figure 45). For many years mathematicians thought that this postulate was not self-evident and that it should be deduced from the other postulates and axioms. However, all their attempts to deduce the fifth postulate failed. We now know that it is impossible to deduce it, which means Euclid's

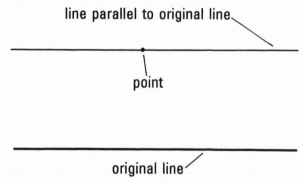

FIGURE 45. Euclid's fifth postulate. Given a line and a point not on that line, one and only one line passes through that point that is parallel to the original line.

fifth postulate has the same status as the continuum hypothesis—both are undecidable. If you want a geometry with the fifth postulate in it, you must include it as an unproved statement (postulate).

Set theory as developed by Cantor is called Naive Set Theory and is based on the intuitive properties of sets. However, modern set theory is based on a collection of axioms developed by two mathematicians, Ernst Zermelo and Adolf Fraenkel. These axioms are now called the Zermelo–Fraenkel axioms or simply ZFC. From ZFC we cannot deduce the truth of the continuum hypothesis. If we want the continuum hypothesis as part of our theory of numbers, we must find an axiom that guarantees it, and add it to the axioms of ZFC. There exist versions of set theory both with and without the continuum hypothesis. Now the question becomes, how does our universe function? Is it better for us to use a set theory where the continuum hypothesis is true or where it is false? Which best suits our needs? At this time we do not know the answer to this question.

The distinction between countable ($\aleph_0$) and uncountable ($\aleph_1$) sets is not trivial. The countable set of numbers, which we usually identify with the natural numbers for convenience, was the first infinite set human beings defined and thought about. In the long struggle to understand extensions in space and their associated magnitudes we have encountered another, different, infinite set, the set $\aleph_1$, which is also designated

countable line segments

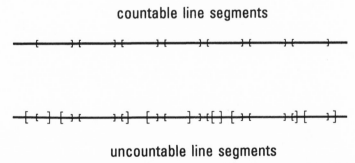

uncountable line segments

FIGURE 46. Countable and uncountable sets. On an infinite line, we can fit an infinite, but countable, number of line segments. However, on the same line, an uncountable number of line segments will not fit, that is, an uncountable number will overlap.

sometimes as c (for continuum). We need this larger set c for our mathematics, and we need our mathematics for our science. A simple demonstration that $\aleph_0$ and c are different may be seen in the following. It is possible to place a countable infinite number of line segments onto a number line without any of them touching each other (Figure 46). But if we try this with an uncountable infinite number of segments, then some of them must overlap. In fact, an uncountable number must overlap.[7]

We have now characterized the entire real number line. We know that it is made up of natural numbers, fractions, algebraic numbers—all of which are countable—and the strange transcendental numbers, which are not countable. What more could we possibly need to know about numbers?

CHAPTER 11

Expanding the Kingdom
Complex Numbers

It may appear at first glance that we have settled the main questions regarding numbers. We have a one-to-one correspondence between points on the number line and real numbers. We are familiar with the different kinds of real numbers, and we know all about countable and uncountable sets, with the strange transcendental numbers a handy example of an uncountable set. What more is there?

When we reviewed Cantor's fine proof that the algebraic numbers are countable, we stumbled over a strange symbol which we intentionally ignored. It was $\sqrt{-1}$, which showed up in both a positive and negative form. What in the world is the square root of a negative one? It must be a number that, when we square it, produces a negative number. But, by the laws of arithmetic, we know that if we multiply two numbers together that have the same sign, we get a positive number for a product. Hence, squaring a negative number produces a positive result as does squaring a positive number. Squaring zero produces only zero. Therefore, it would seem that no real number on our number line will square to give us a negative product.

We may be tempted to just sweep our $\sqrt{-1}$ under the rug and forget it. This is what many mathematicians did for centuries. However, the problem did not go away. We ran across a positive and a negative $\sqrt{-1}$ as two solutions to the polynomial $x^2 + 1 = 0$. This is a fairly simple polynomial, and both of its solutions are this strange square root of a negative number, which we will call a negative radical. The polynomial will have no solution if we do not somehow account for $\sqrt{-1}$. To make the situation worse, this kind of solution has many forms. The polynomial $x^2 + 2 = 0$ has two solutions, $\sqrt{-2}$ and $-\sqrt{-2}$. Now we must find a number that when squared is a negative two! We suddenly realize that for every real number, r, there must be a square root of a negative r because of the polynomial $x^2 + r = 0$. To solve this general polynomial, we must take r to the right side of the equation, making it negative, and then find the square root of this number. This implies that there are as many of these strange negative radicals as there are real numbers! To make matters even more complex we have the polynomial $x^4 + 1 = 0$. We solve this equation to find that $x = \sqrt[4]{-1}$, that is, a number which, when multiplied by itself three times, produces a negative number. Hence, we have other radicals besides the square root involving these strange numbers. Where the heck are they, if they are not on the real number line? Do they really exist?

For many centuries, the radicals of negative numbers were simply ignored by mathematicians. If a polynomial had only negative radicals for solutions, then the mathematicians declared that the polynomial was unsolvable. The greatest Greek algebraist, Diophantus, soundly rejected negative radicals as roots.[1] One of the first to even consider the square roots of negative numbers as solutions to polynomials was Albert Girard (1590–1633), yet few others of the seventeenth century would do so.[2] It would take a major revamping of both algebra and geometry before these strange entities could find a secure mathematical home. This would occur through the efforts of two men who have already been mentioned, René Descartes and Carl Gauss. Descartes would provide the necessary tools, while Gauss would give the final solution.

THE BIRTH OF ANALYTIC GEOMETRY

During his lifetime, René Descartes (Figure 47) never realized that his work would lay the foundation for an entire new class of numbers. In fact, he rejected the square roots of negative numbers as possible

FIGURE 47. René Descartes, 1596–1650. (Photograph from Brown Brothers, Sterling, PA.)

solutions to polynomials and even coined the modern name of *imaginary numbers* for such roots.

Descartes was born March 31, 1596, in La Haye, France, to a noble French family.[3] Wealthy enough not to worry about working, he became a seventeenth-century-gentleman with a taste for gambling and women, yet he possessed a fine classical education. During his youth he amused himself by serving as a soldier and fighting for various European princes. All this ignores the fact that he was a first-rate mathematician and philosopher, and even found time to contribute to the sciences. He is best known for his famous argument for existence, "I think, therefore I am." His single most significant contribution is the subject we deal with here: analytic geometry.

Supposedly, at the age of twenty-three and while in the service of the Bavarian army, Descartes had three traumatic dreams, which he interpreted as a sign to give up his unproductive lifestyle and contribute to mathematics, philosophy, and science. He reported receiving a magic key in his dreams that would open up a great wealth of truth for him.[4] He never divulged this magic key but his biographers presume it was the basis of his geometry.

During the classical Greek period, the discovery of the incommensurability of the diagonal (i.e., that $\sqrt{2}$ is irrational and cannot be represented as a fraction) caused the fields of algebra and geometry to go their separate ways. Algebra dealt with numbers, while geometry dealt with extensions in space. Descartes was able to pull these two disciplines back together again. He did this with a simple construction. First he took a real number line (although at this early date, he did not fully appreciate how rich this line was) and placed it horizontally. He then took a second number line and drew it vertically so that it intersected the first line in a right angle (Figure 48). Each line is called an *axis*, and where the lines meet is the *origin*, which is assigned the value of zero. Now, every point in the plane can be located by giving two numbers: its distance along the horizontal axis from the origin and its distance along the vertical axis from the origin. For brevity, the horizontal axis is called the *x-axis*, while the vertical is called the *y-axis*.

For example, Figure 48 shows a point above the number 2 on the x-axis and to the right of the number 3 on the y-axis. This single point is

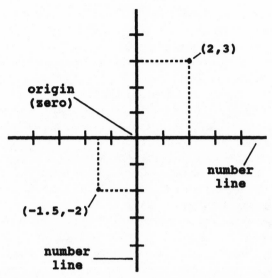

FIGURE 48. Cartesian coordinate system. Every point in the plane is uniquely defined by two real numbers (x,y) where x is the horizontal distance from the origin and y is the vertical distance from the origin.

identified with the two numbers $(2,3)$. In this manner every point is identified with a set of two numbers. The second point in Figure 48 is located at $(-1.5, -2)$ which means we go left of the origin 1.5 units and down from the origin 2 units to locate the point. The two numbers associated with each point in the plane are called the point's *coordinates*. In honor of Descartes, we call this system the Cartesian coordinate system.

Descartes did much more than just invent this fine system to locate points on a plane as number sets, he also realized its utility in describing geometric curves as algebraic relationships. This is the key to analytic geometry. The algebraic equation for a circle whose center is at the origin is $x^2 + y^2 = r$, where x and y represent the coordinates of the points on the circle and r is the radius. Though the circle had been well studied in ancient times, analytic geometry took us much further. It gave us the power to deal with curves, which had been difficult to handle in Euclidean geometry. The equation for an ellipse, an elongated circle that describes the orbits of planets, is $x^2/a^2 + y^2/b^2 = 1$ where a and b are the major and minor axes. A parabola is the set of points a fixed distance from a point and a line. Its equation is $y^2 = 4mx$ where m is the distance of the fixed point from the origin.

Curves such as the parabola and ellipse help describe many natural phenomena that we observe around us. Being able to represent these complex geometric curves as algebraic relationships made modern geometry possible. As we will see, Descartes' new system of point representation also made it possible to understand the negative radicals. However, Descartes, in his day, was not ready to take this step. The use of Descartes' coordinate system to understand them had to wait almost two hundred years for the work of Carl Gauss.

Descartes stumbled onto his key to analytic geometry in 1619, yet he did not publish the results until 1637—a delay of eighteen years. During that time he fought in several campaigns and had other close calls to his well-being. As E. T. Bell has pointed out in his *Men of Mathematics*, a stray musket ball could have ended Descartes's career long before 1637, and the world would have had to wait for his greatest discovery. The last years of Descartes life were devoted to his philosophy, mathematics, and science. In 1650 he fell ill and died at the age of 53.

GAUSS LOCATES THE COMPLEX NUMBERS

Carl Friedrich Gauss was born on April 30, 1777, to humble peasant parents. He was a child prodigy, and his mother, Dorothea Benze Gauss, pushed her son toward an education to break free from his peasant background. This he did with the help of the Duke of Brunswick, who gave Gauss an allowance that enabled him to enter the Caroline College at age fifteen and the University of Göttingen at eighteen. He received his doctorate from the University of Helmstädt in 1799 at the tender age of twenty-two. Unable to secure a mathematical post to his liking, he became Director of the Göttingen University Observatory and, later, Professor of Astronomy. Yet, mathematics remained his one obsessive love. Through his mathematical discoveries he quickly developed a reputation throughout Europe as one of the world's greatest mathematicians, yet his personal life was punctuated with numerous tragedies, including the death (due to illness) of two wives (the first after giving birth) and a child.[6] This caused Gauss to dread death and, in his later years, to have a great mistrust of doctors.

Gauss published several outstanding works, contributing to the fields of number theory, complex analysis, differential geometry, topology, the application of mathematics to astronomy, and the sciences of magnetism, crystallography, optics, and electricity. His personal notebook, which was not published for almost half a century after his death in 1855, revealed he had also anticipated non-Euclidean geometry. Yet, it is the consequences of his doctoral thesis that concerns us here. As previously mentioned, his thesis of 1799 proved for the first time that every polynomial has at least one solution. However, to prove this, Gauss had to come up with a sound foundation to the problem of radicals of negative numbers. Since the polynomial $x^2 + 1 = 0$ has only radicals of negative numbers as solutions, the Fundamental Theorem of Algebra would be false if we had no way to deal with them.

Gauss used Descartes's system of two intersecting number lines. The horizontal x-axis he designated as the real number line, which we are familiar with. The vertical axis became another number line of imaginary numbers based on $\sqrt{-1}$. For convenience, we frequently substitute the letter i for $\sqrt{-1}$. Every point on the x-axis is a real number. Every point on the y-axis is a real number multiplied by i or $\sqrt{-1}$. From this, we see that $\sqrt{-1}$ is one unit up from the origin on the y-axis (Figure 49), while $2\sqrt{-1}$ is the point two units up from the origin on the y-axis, and $-\sqrt{-1}$ is simply one unit down from the origin on the y-axis. What are all the points in the plane that are not on the x-axis or the y-axis? They are the complex numbers which are a combination of real numbers and imaginary numbers. Each point is represented by two numbers, a and b, and these two numbers define the complex number $a + bi$, where a is the real part and b is the imaginary part.

In this manner Gauss defined an expanded class of numbers, the complex numbers, where the real numbers are a subset of the complex numbers. The real numbers are those complex numbers that have an imaginary part equal to zero. Conversely, the y-axis represents those

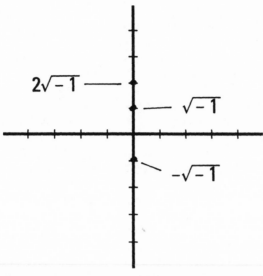

FIGURE 49. The Gaussian plane. The values on the horizontal axis represent real numbers while the numbers on the vertical axis represent multiples of $\sqrt{-1}$. Each point in the Gaussian plane is represented by two real numbers in either of two forms: (a,b) or $a + bi$ where $i = \sqrt{-1}$. Frequently called the complex plane.

purely imaginary numbers that are complex numbers with real parts equal to zero. To multiply the real number, 1, by $\sqrt{-1}$ rotates that number counterclockwise from the x-axis up to the corresponding position of $\sqrt{-1}$ on the y-axis. If we multiply it again by $\sqrt{-1}$ it rotates the number counterclockwise again down to the corresponding negative part of the x-axis (Figure 50). This second rotation gives us $\sqrt{-1}\cdot\sqrt{-1} = -1$. Hence, multiplying 1 by $\sqrt{-1}$ twice produces the -1 we are after. In shorthand we have $(\sqrt{-1})\cdot(\sqrt{-1}) = i\cdot i = i^2 = -1$.

It is now necessary to ensure that the normal operations of arithmetic make sense with complex numbers. To add two complex numbers we have:

$$(a + bi) + (c + di) = a + bi + c + di = (a + c) + (b + d)i$$

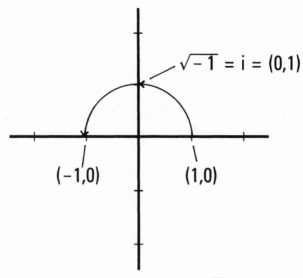

FIGURE 50. The Gaussian plane. Multiplying 1 by $\sqrt{-1}$ is equivalent to rotating 1 counterclockwise 90° to the vertical axis to get $\sqrt{-1}$. Multiplying $\sqrt{-1}$ by $\sqrt{-1}$ rotates the product 90° counterclockwise again to the horizontal axis where it has a value of -1. Hence, $\sqrt{-1} \cdot \sqrt{-1} = -1$.

Therefore, to add two complex numbers we simply add the real parts (the a and c) and the two imaginary parts (the b and d). Subtraction is just as easy:

$$(a + bi) - (c + di) = a + bi - c - di = (a - c) + (b - d)i$$

Here again, we simply subtract the corresponding real parts and then the corresponding imaginary parts. Multiplication is interesting for we get

$$(a + bi) \cdot (c + di) = ac + adi + bci + bdi^2$$

However, $i^2 = -1$, so we can substitute -1 into the last term to get

$$(a + bi) \cdot (c + di) = ac + adi + bci - bd = (ac - bd) + (ad + bc)i$$

The division of two complex numbers is considerably more difficult:

$$(a + bi)/(c + di) = [(ac + bd) + (bc - ad)i]/(c^2 + d^2)$$

Yet, when we are done with the four operations, we always get other complex numbers. Therefore, the complex numbers are closed under the four operations of arithmetic, except for the special case of division by zero. In addition, complex numbers can be raised to powers and have their radicals computed, and the resulting numbers are still complex numbers. Hence, under these two additional operations, the complex numbers are closed.

Solving for polynomials is now greatly enhanced. If we consider the polynomial $x^2 - 6x + 11 = 0$, we get the two solutions, $3 + \sqrt{2}i$ and $3 - \sqrt{2}i$. To locate these two solution points on the plane of complex numbers (Gaussian plane) we move to the right of the origin on the x-axis three units (Figure 51). Next we move up a distance of $\sqrt{2}$ (approximately

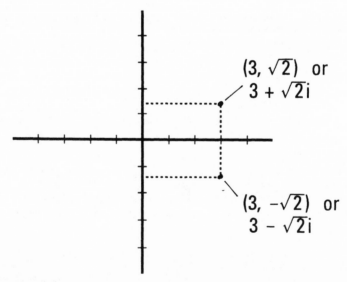

FIGURE 51. Locating the two solutions to the polynomial, $x^2 - 6x + 11 = 0$, on the Gaussian plane.

1.414 units) above the origin and locate our first point, $3 + \sqrt{2}i$. For the second point we again move along the x-axis three units, then we move a distance of $\sqrt{2}$ down to locate $3 - \sqrt{2}i$. In fact, we find that the complex numbers that have imaginary parts will occur in pairs of the form $a + bi$ and $a - bi$. Such pairs are called complex conjugates.

Carl Gauss finally attempted what the other mathematicians had avoided. He claimed an existence for imaginary numbers (in fact all complex numbers) that was just as objective as the real numbers. If negative numbers and irrational numbers existed, then the complex numbers did too. Surprisingly, while Gauss was defining complex numbers as points on a plane, two other mathematicians were attempting to do the same thing. One was the self-taught Norwegian mathematician Caspar Wessel (1745–1822), and the second was the Swiss bookkeeper, Jean Robert Argand (1768–1822). Both these men suggested a geographical means of representing complex numbers, but their work was generally ignored.

HOW MANY ARE THERE?

We have obviously taken a huge step in expanding our concept of numbers. Looking at the real number line embedded in the Gaussian plane, we may have the strong intuitive feeling that there are many more points on the plane than on any single line. This same intuitive feeling was shared by most mathematicians. In fact, through each point on the y-axis we can draw a straight line that is parallel to the x-axis. This line contains as many points as the real number line. Since there are an uncountable number of points on the y-axis, we can construct an uncountable number of lines, each containing an uncountable number of points. Surely, this uncountable set of uncountables is, itself, far larger than the set of points on a single line.

To answer this question we must once again turn to Georg Cantor. And again we find the question raised in a letter from Cantor to his friend, Richard Dedekind, written on January 5, 1874.

> Is it possible to map uniquely a surface (suppose a square including its boundaries) onto a line (suppose a straight line including its end

points), so that to each point of the surface one point of the line and reciprocally to each point of the line one point of the surface corresponds?[7]

This time it took Cantor more than eight days to come up with the solution. It was three years later, in 1877, that he sent Dedekind the outline of a solution. What Cantor did was demonstrate a way to produce a one-to-one mapping between the points on a plane and a single line. Each point on the plane is represented by two numbers (a,b) that represent the real and imaginary parts of a complex number, $a + bi$. Both of these numbers have a unique decimal expansion. For convenience we will assume both a and b are between zero and 1.

$$a = 0.a_1a_2a_3a_4 \ldots \text{ and}$$
$$b = 0.b_1b_2b_3b_4. \ldots$$

We can use these two decimal expansions to create a unique real number on the real number line. We do this by alternately choosing the digits from a and b to construct this number, p.

$$p = 0.a_1b_1a_2b_2a_3b_3a_4b_4 \ldots$$

The constructed number p is a real number uniquely mapped onto the complex number $a + bi$. This kind of construction produces a one-to-one mapping of all complex numbers from the complex plane onto the real number line. This proves that the cardinal number for the real number line, $\aleph_1$, is also the cardinal number for all the complex numbers in the complex plane. From this we realize how very rich and dense the number line is in numbers, since the "number" of points on the line is the same as the "number" of points in a plane. This certainly goes against our intuitive feeling, as it did with Cantor. Once he had his proof in hand, even he had a hard time accepting it.[8]

Yet, significant differences do exist between the complex plane and the real number line. All numbers on the number line are ordered by the relation of "greater than." That is, if a and b are numbers, and a is not equal to b, then either $a > b$ or $b > a$. In other words, the two numbers are ordered according to size. This is not true on the complex plane. We cannot say in a consistent fashion that, given two complex numbers,

$a + bi$ and $c + di$, one is larger than the other. Therefore, while the complex numbers form a number *field*, this field is not ordered. A number field is any set of real or complex numbers such that the sum, difference, product, and quotient (except for zero) is another number in the set. Therefore a number field is closed under the four operations of arithmetic. The real numbers form an *ordered number field*. As with the real numbers, the complex numbers are either rational (when both a and b in $a + bi$ are rational) or irrational, and they are either algebraic or transcendental (when either a or b is transcendental). Picturing the complex plane and realizing that the "bulk" of numbers on both the x-axis and the y-axis are transcendental, makes us appreciate how richly populated the plane is with complex transcendental numbers.

The definition of complex numbers did more than just tidy up a nuisance in algebra. It opened an entirely new branch of mathematics known as complex analysis. This has allowed mathematicians to make a more powerful penetration into the secrets of theoretical mathematics and to define a greater range of models in applied mathematics. Complex numbers do more than satisfy polynomials, for they explain certain other expressions that make no sense if we are limited to only the real numbers. For example, the equation $e^x = -1$ has no real number solution. However, if we allow for complex numbers, the solution is πi or $\pi\sqrt{-1}$. This complex number point is located on the y-axis a distance of π above the origin. In fact the identity of $e^{\pi i} + 1 = 0$ was first recognized by Leonhard Euler in 1748 long before Gauss developed a sound foundation to complex numbers. This identity is considered by mathematicians the world over as one of the most profound and beautiful relationships in all of mathematics. It incorporates into one equation the operation of addition, the relationship of equality, the primary numbers of 1 and 0, both transcendental numbers e and π, and the complex number i or $\sqrt{-1}$. All this rolled up into such a simple expression dazzles the mind.

HAMILTON'S QUATERNIONS

Once Gauss expanded numbers from the real line to the complex plane, it was only a matter of time before someone considered whether the process could be continued into three-dimensional space. That is, did a set of numbers exist as number triplets (a, b, c) that correspond to points in three-dimensional space, and also act like numbers? This someone was William Rowan Hamilton, perhaps Ireland's greatest mathematician. He was born in Dublin in 1805 and proved to be a precocious child, learning numerous foreign languages before his teens from a polyglot uncle. In later years he claimed that by thirteen he knew thirteen languages. Fortunately, the young Irishman satisfied his linguistic curiosity and turned his attention toward mathematics.

At the age of seventeen Hamilton entered Dublin's Trinity College. Five years later he was still a student when he was appointed Astronomer Royal of Ireland and a professor of astronomy at Trinity.[9]

In 1828 Hamilton began to consider expanding the idea of complex numbers into three-dimensional space. He needed a way to define the operations of addition and multiplication of his triplets such that the basic laws of algebra were not violated. These laws include the following rules for any three numbers a, b, and c.

1. $a + b = b + a$ (commutative law of addition)
2. $a \cdot b = b \cdot a$ (commutative law of multiplication)
3. $(a + b) + c = a + (b + c)$ (associative law of addition)
4. $(a \cdot b) \cdot c = a \cdot (b \cdot c)$ (associative law of multiplication)
5. $a \cdot (b + c) = a \cdot b + a \cdot c$ (distributive law)

However, Hamilton could not find the proper definitions that would preserve the above five laws. The multiplication of a complex number in the plane was defined as a kind of rotation. Yet, when Hamilton tried to define multiplication in three dimensions as a rotation, he failed. As it turns out, using triplets to define a rotation in three dimensions is not possible. What about four dimensions? Could numbers be defined in four dimensions that preserved the laws of algebra? Hamilton's search moved to consider a kind of "super-complex" number of the form $a + bi + cj + dk$.

For fifteen years Hamilton struggled with this problem. Finally, the solution came as a flash while he was walking with his wife. The date was October 16, 1843, and once again we know the exact day of a great mathematical breakthrough. Hamilton realized he had to give up one of the laws of algebra, the commutative law of multiplication. As long as he did not insist that $a \cdot b = b \cdot a$, he could define the four arithmetic operations for his new numbers, which he named *quaternions*. These numbers were defined as real number quadruples with the form $a + bi + cj + dk$ where the following identities hold:

$$i^2 = j^2 = k^2 = -1 \text{ and } ijk = -1.$$

Hamilton succeeded in defining the operations such that all the laws were true except the commutative law of multiplication. Today we know that only algebras defined up to two dimensions can keep the commutative law of multiplication. Any algebras defined for higher spaces must give up this law.

Hamilton considered these quaternions as his greatest contribution and spent the majority of his life expounding on them. Even though modern mathematics no longer uses quaternions as such, their development did lead to several significant advances. A quaternion has two parts, the purely real part, a, which is called the *scalar*, and the "imaginary" part, $bi + cj + dk$, which is called the *vector*. The vector part is a kind of directional line in three dimensions. It is used in many fields of science today and its study is known, mathematically, as vector analysis.

Hamilton's quaternions made algebraists realize that new algebras could be developed by considering changes in the basic laws of algebra. This is analogous to the development of non-Euclidean geometries by using alternatives to Euclid's fifth postulate. Hence, the theory of quaternions pointed to the beginning of modern abstract algebra.

Before we part from Hamilton's quaternions, we must ask one of our favorite questions: Just how many quaternions are there? Here again, Cantor provided us with the answer. We will repeat the same procedure to map the quaternions of hyperspace into a real line. We write the four numbers constituting a specific quaternion in their decimal expansions. For convenience we again assume they are numbers between zero and one. Hence we get

$$a = 0.a_1 a_2 a_3 a_4 \ldots$$
$$b = 0.b_1 b_2 b_3 b_4 \ldots$$
$$c = 0.c_1 c_2 c_3 c_4 \ldots$$
$$d = 0.d_1 d_2 d_3 d_4 \ldots$$

From these four numbers, which define our specific quaternion, we build the corresponding real number, p.

$$p = 0.a_1 b_1 c_1 d_1 a_2 b_2 c_2 d_2 a_3 b_3 c_3 d_3 a_4 b_4 c_4 d_4 \ldots$$

Thus from all the countless quaternions in four-dimensional space we get a one-to-one mapping onto the real numbers. The cardinal number for the quaternions, like the complex numbers, is the same as the cardinal number for the real numbers. This means there are as many points on a single line as in a plane, three-dimensional space, and even four-dimensional space. In fact, the "number" of points is not a determining factor in dimensionality. Cantor proved that for a countable number of dimensions, the cardinal number of points is equal to the cardinal number for the line, or $\aleph_1$. This is difficult to accept intuitively, and many mathematicians hung on to the belief that Georg Cantor must have made a mistake.

We have now considered those numbers we use in our daily activities, and in addition, we have looked at the complex numbers, numbers essential in the everyday workings of science. In our next step we shall look at strange and exotic numbers that ninety-nine percent of the general population have never even heard of. Prepare yourself for an adventure.

CHAPTER 12

Really Big
Transfinite Numbers

DEFINING TRANSFINITE NUMBERS

In our study of transcendental numbers, we reviewed how Cantor defined two infinite cardinal numbers, $\aleph_0$ and $\aleph_1$, for two kinds of infinite sets, sets that are countable and sets that are uncountable. If N is the infinite set of all natural numbers, then we define $\overline{N}$ as the cardinal number of N. In this way we avoid confusing a set with its cardinal number. In a similar fashion we let R stand for the set of all real numbers. Then $\overline{R}$ is its cardinal number and is frequently shown as simply c. We know that $\overline{N} = \aleph_0$. Cantor believed that $c = \aleph_1$ but could not prove it. We also know that the question has been asked whether there are any infinite cardinals between $\aleph_0$ and c, and the answer is undecidable.

However, Cantor went far beyond these first two infinite cardinal numbers, for he actually defined an infinite sequence of such numbers with $\aleph_0$ and $\aleph_1$ being the first two. These cardinal numbers became the *transfinite numbers*—numbers beyond or larger than the finite numbers. The method of generating each successive transfinite number is quite simple. If we have a finite set with n elements, then we investigate how many possible ways we can group these n elements. All these possible groupings become elements to a new set with a higher cardinal number than the original set. Yet, when we consider all the kinds of sets possible, and especially the infinite sets, how do we know we can form all the possible combinations of elements? For that we have an axiom from set theory:

AXIOM OF POWERS: For each set there exists a collection of sets that contains among its elements all the subsets of the given set.

This axiom guarantees that we can always form our set of element combinations. This axiom is obvious with finite sets but is not so obvious when we are dealing with infinite sets.

For example, consider the set with the three elements 1, 2, 3. We show this set as $\{1, 2, 3\}$. How many different ways can we show these three elements? The answer turns out to be eight, and the eight different sets become $\{\varnothing\}$, $\{1\}$, $\{2\}$, $\{3\}$, $\{1,2\}$, $\{1,3\}$, $\{2,3\}$, $\{1,2,3\}$. The first set, $\{\varnothing\}$, is called the empty set and is frequently written as just $\varnothing$, without the brackets. The empty set is one possible way to show the three elements, that is, not to show any of them. Now the set containing our eight sets is $\{\varnothing, \{1\}, \{2\}, \{3\}, \{1,2\}, \{1,3\}, \{2,3\}, \{1,2,3\}\}$ and has a cardinal number of 8. The formula that tells us a set with three elements can form eight different sets is $2^3 = 8$. We can generalize this: A set with n elements can be formed into 2^n different sets. The number of sets we can form from n is always larger than n. Hence, the finite cardinal number for 2^n is greater than n. Cantor, in his famous letter to Dedekind in December of 1873, proved this very relationship for infinite sets; that is, if n is the cardinal number for an infinite set, then $n < 2^n$. This is now called Cantor's theorem.

CANTOR'S THEOREM: For every cardinal number n, $n < 2^n$.

If we begin with $\aleph_0$ we can actually generate the sequence of infinite cardinal numbers by taking successive powers of 2 in the following manner: $\aleph_1 = 2^{\aleph_0}$. In a similar fashion, $\aleph_2 = 2^{\aleph_1}$, and $\aleph_3 = 2^{\aleph_2}$. Each time we simply raise 2 to the current transfinite number to generate the next transfinite number. In this way we generate an infinite set of transfinite numbers: $\aleph_0$, $\aleph_1$, $\aleph_2$, $\aleph_3$, $\aleph_4$, $\aleph_5$, $\aleph_6$, $\aleph_7$, ...

By using Cantor's theorem we can visualize how $\aleph_0$ generates c. We will prove that $2^{\aleph_0} = 10^{\aleph_0}$. Now $\aleph_0$ is just the countable set $\{1, 2, 3, . . .\}$, so that $10^{\aleph_0}$ represents every possible way that each digit in each set can be grouped into ten digits.[1] Each individual combination yields a unique set $\{a, b, c, \ldots\}$ that can represent a decimal expansion of a number between 0 and 1. Therefore, all combinations represented by $10^{\aleph_0}$ should include every decimal expansion between zero and 1, and the cardinal number of this set is c. But c is also the cardinal for the entire real line, in addition to being the cardinal number for points between zero and 1. Therefore, we have $2^{\aleph_0} = 10^{\aleph_0} = c$. Another strange result is that $\aleph_0! = 1 \cdot 2 \cdot 3 \cdot 4 \ldots = c$.

As numbers, transfinite numbers must obey a set of laws just as finite numbers do. For example, we have addition and multiplication defined for transfinite numbers, but they are rather boring. For example, if $v < w$ then $\aleph_v + \aleph_w = \aleph_w$, and $\aleph_v \cdot \aleph_w = \aleph_w$. Hence, the cardinal number of the sum and product is just the largest cardinal number of the two factors. This means that $\aleph_v^2 = \aleph_v$, and, in fact, $\aleph_v^n = \aleph_v$ for any finite n. It also means that $\aleph_v + \aleph_v = 2\aleph_v = n \cdot \aleph_v = \aleph_v$ for every finite n. Hence, transfinite numbers do not do much when added or multiplied together. It's when they are used as exponents that we get interesting results, such as the production of new transfinite numbers.

In addition to a theory for transfinite cardinal numbers, Cantor developed a parallel theory for ordinal numbers. From our definitions of ordinal and cardinal numbers we will remember that a cardinal number is associated with the manyness of a set, while an ordinal number is associated with a sequence of numbers that progresses, one step at a time, up to the last ordinal number. The natural number sequence $\{1, 2, 3, 4, 5, \ldots\}$ is just such a sequence. The upper limit we achieve by adding 1 to each ordinal number to generate the next is called ω (small omega), and we have $\omega = \aleph_0$. This says that the first transfinite ordinal number, ω, is equal to the first transfinite cardinal number. Cantor then went on to define larger ordinal numbers corresponding to his cardinal numbers.

It's difficult to grasp how much of an increase we get when we move from each transfinite number to its successor. When we move from $\aleph_0$ to $\aleph_1$, we move from the countable set of the natural numbers to the uncountable set of all real numbers, and the real numbers are sufficiently dense or populous to fill up the space of the entire universe. Since $\aleph_1$ is uncountable, then all the larger transfinite numbers are also uncountable, which makes $\aleph_1$ our first uncountable cardinal. How much more "multiplicity" do we encounter when we move from $\aleph_1$ to $\aleph_2$? Are there any examples of $\aleph_2$? Mathematicians say that $\aleph_2$ is equal to the cardinal number of the set of all real functions over the real line interval between 0 and 1. This may be a mathematical truth, but it does not convey to our minds the immense richness that $\aleph_2$ must represent. And what of $\aleph_3$ and $\aleph_{10}$ or even $\aleph_{1,000,000}$? These are certainly numbers so large that we have no way to grasp their multiplicity. Yet we are just beginning!

AND NOW FOR SOME REALLY BIG NUMBERS

We can keep defining new, bigger transfinite numbers. We can run through all the natural numbers as subscripts for $\aleph$, ending with ω to get $\aleph_\omega$, which is really the same as $\aleph_{\aleph_0}$. We can continue in the same manner and attach bigger subscripts, and subscripts of subscripts, but it will all be rather boring. We will end up with just a bunch of symbols stuck together without any additional appreciation of an expanded many-ness. To do this, we must first look at the *Absolute Infinite*.

Before Cantor, philosophers and mathematicians generally defined only two categories: The finite and the infinite. Individual things belonged to the finite and could be grasped by the intellect. On the other hand, the infinite was not comprehensible and was beyond reason, and was frequently identified with God. This became the Absolute Infinite. As Rudy Rucker points out in *Infinity and the Mind*,

> In terms of rational thoughts, the Absolute is unthinkable. There is no noncircular way to reach it from below. Any real knowledge of the Absolute must be mystical, if indeed such a thing as mystical knowledge is possible . . . . Even if full knowledge of the Absolute is only possible through mysticism, it is still possible and worthwhile to discuss *partial* knowledge of the Absolute rationally.[2]

Cantor believed in the Absolute Infinite and identified it with God.[3] His thesis was that there were rational infinities below this Absolute Infinite that we can talk about. In fact, he believed that his theory of transfinite numbers came to him from God. The Absolute Infinite has also been identified with the collection of all infinities. We cannot call such a collection a set, for a set has a specific, and hence, limiting definition. A set is a manyness that can be conceptualized as a unity. The Absolute cannot be grasped by the mind as a unity. We designate the Absolute as Ω. Can Ω be an ordinal number? No, because if it were, then we could add 1 to generate a bigger ordinal, and there is nothing greater than Ω. In similar fashion we can say the Ω is not simply a cardinal number, because if it were, then we could form the next cardinal number as 2^Ω, and this would also lead to a contradiction since there would then be something bigger than Ω. Therefore, Ω is not an ordinal or a cardinal because it is beyond all ordinals and cardinals and is, as a unity, unknowable. However, this will not exclude the possibility that Ω does have some properties it shares with the transfinite numbers.

Now that we have defined Ω (which we really cannot do in practice) as that which is beyond all transfinite infinities, we can go on to define some rather more interesting transfinite numbers. Two properties of Ω will help us. First, for every property that Ω possesses, then some transfinite number must also possess this property. Why? If Ω has a unique property, p, which is shared by no other infinity, then we could uniquely describe Ω as that infinity with property p. But then Ω would not be Absolute and beyond definition. Hence, for every property that Ω has, at least one transfinite number also has it. Notice from this same argument that we get a much stronger statement, for if Ω was only one of two infinities with property p, then we could still get a unique statement defining Ω, such as Ω is the larger infinity with property p. Since we cannot even do this, we can say that there are infinitely many transfinite numbers with property p. This property of Ω, that it must share all its properties with other transfinite numbers, is called the *Reflection Principle*.[4] That is, Ω reflects its own properties down to the transfinite numbers.

The second interesting property of Ω is that we can never reach Ω by building larger and larger transfinite numbers. Strictly speaking, we say that Ω cannot be reached by building up constructions of numbers less than Ω. Were this not true, we could define Ω as the successive construction of certain transfinites. Hence, we say that Ω cannot be reached from below, or that it is *inaccessible*. Now, because of the Reflection Principle, we know that an infinite number of transfinite numbers must also share the property of inaccessibility. This becomes our new class of gargantuan numbers: *inaccessible transfinite numbers*. No matter how hard we try, we can never construct these numbers by building up ever larger combinations of $\aleph$. They are just too big—they are inaccessible.

Technically, $\aleph_0$ is our first inaccessible transfinite number because we cannot reach it by adding more and more of the finite numbers below it. It is called the zeroth inaccessible number because we usually start counting with zero in set theory (hence the 0 in $\aleph_0$). But the successive $\aleph$ we have talked about are built from below beginning with $\aleph_0$, and are therefore not inaccessible. The first inaccessible transfinite beyond $\aleph_0$ is called θ (Greek theta). This θ is called the first large cardinal among the transfinite numbers.

How many inaccessible numbers are there beyond θ? We can build an argument that there must be Ω inaccessible transfinites. We could start with θ_1 and then define the second θ_2 and its successor, and in this fashion build up a whole sequence of inaccessible transfinites. But we can do much better. We will simply define an inaccessible transfinite as being inaccessible, and its degree of inaccessibility is not reachable from below beginning with θ. It is so inaccessible that its position among the inaccessibles is inaccessible. It is called a *hyperinaccessible* number. We are now really getting up into the stratosphere of big numbers. Of course, we could go on using the same principles and define "hyper-hyperinaccessibles," and then on to "super-duper-hyper-hyper," and so forth, but it then becomes boring.

However, transfinite numbers have been around so long that mathematicians have busily defined ever larger transfinite cardinal numbers. Beyond the hyperinaccessibles are the Mahlo cardinals, indescribable cardinals, ineffable cardinals, partition cardinals, Ramsey cardinals, measurable cardinals, strongly compact cardinals, supercompact cardinals, and the extendable cardinals. The extendable cardinals are the largest, and some mathematicians claim they do not even exist, while others say they are the largest we will ever find.[5] What is important here is that each new and larger cardinal is not just a simple successor from the previous cardinal, but represents a new and higher platform toward the Absolute Infinite. Perhaps what we have here is analogous to the situation in physics: splitting the atom, and then the subatomic particles, and always finding there is a further subdivision and more elementary particles. Perhaps we will continue defining larger cardinals as long as we have a desire to do so.

We can summarize the transfinite numbers in the following, even if incomplete, Table 11. We assume it is incomplete because just as soon as we declare it perfect, some mathematician will prove us wrong.

CANTOR'S PERSONAL STRUGGLE FOR THE TRANSFINITE

Cantor's proof of the uncountability of the real numbers was published in 1874. Yet, his ideas and theories were not universally accepted. Indeed, Cantor's work ignited a bonfire of controversy over the concept of infinity. As mentioned earlier the major antagonist to Cantor's ideas was one of his professors at the University of Berlin, Leopold Kronecker. He was the one who coined the pithy phrase, "God made the integers, the rest is the work of man." Kronecker belonged to the modern school of

TABLE 11. Partial List of Transfinite Numbers

Name	Symbol(s)	Comments
Aleph-null	$\aleph_0$	$= \omega$, number of natural numbers, fractions, algebraic numbers.
Aleph-one	$\aleph_1$	$= (?)$ c number of points on the real number line.
Alephs	$\aleph_0, \aleph_1, \aleph_2, \aleph_3$	The sequence of alephs built up from $\aleph_0$.
Inaccessible	θ	Numbers too big to be reached from below, and depending on the Reflection Principle of Ω.
Hyperinaccessibles		Inaccessibles too big to be reached from θ_1.
Large cardinals: Mahlo, indescribable, ineffable, partition, Ramsey, measurable, strongly and supercompact, extendable		Additional kinds of transfinite numbers that cannot be reached from below.
Absolute Infinity	Ω	Beyond all number, collection of all sets, God to some.

intuitionism, which rejected the infinite as a completed thing and considered it only as a potential. This is exactly the stand expounded by Aristotle some twenty-two centuries before. In this sense, Kronecker was a modern in the camp of Pythagoras, Plato, and Aristotle. This resistance to the infinite extended to irrational and transcendental numbers in addition to the transfinite numbers. Surprisingly, both Kronecker and Cantor were religious men, and Platonists. However, Kronecker felt that transfinite numbers detracted from God's infinite greatness, since only God could have the attribute of being infinite. If anything else had this attribute, it would diminish God's greatness. However, Cantor believed the transfinite numbers added to God's greatness. The transfinite numbers were unimaginably large, and God's infinite nature was even beyond them.

And Kronecker was not alone in his opposition to transfinite numbers, for other mathematicians also believed that allowing infinite processing into mathematics caused great harm to the security of its sound logical foundation.

Cantor's disagreements with Kronecker had disastrous results for Cantor. Kronecker not only tried to delay or deny publication of some of Cantor's early papers, but he used his position and prestige at the University of Berlin to attack forcefully Cantor's theories. It was an unbalanced fight. Kronecker was already an established mathematician who had made his mark. He was considered one of Germany's premier mathematicians. Cantor, on the other hand, was a young, unproven mathematician at a minor university. Kronecker's attacks became so intense that other mathematicians took notice and began to express sympathy for Cantor.[6] Yet, Kronecker did not single out Cantor alone but was willing to enthusiastically denounce anyone who advocated the idea of a mathematics of the infinite.

In May of 1884, Cantor suffered a complete nervous breakdown that lasted for approximately a month. He attributed this illness to overwork on the Continuum Hypothesis (is $\aleph_1 = c$?) and sensitivity to the attacks from Kronecker. Although he seemed to have fully recovered when he returned to work, in the following years the attacks of mental instability recurred with increasing intensity. Cantor seemed to suffer from some kind of deep depression. In 1899 his youngest son died. This in conjunction with other stressful events hospitalized him in the Halle Nervenklinik. His depression grew more intense, and in the 1902–1903 winter term he was relieved of his teaching duties and hospitalized.

From this period to the end of his life in 1918, Cantor was in and out of hospitals. It has been reported by his biographers that it was Kronecker's attacks that drove him to mental instability.[7] This hypothesis is certainly questionable, for we will probably never know what physiological factors, if any, contributed to his illness. By some standards, Cantor achieved a successful married life with his wife, Vally, and fathered six children. Therefore, for a portion of his life he was able to function successfully. While his mathematical views were under constant attack by Kronecker, other well-known mathematicians supported him, encouraging his work. Kronecker died in 1891, yet Cantor's illness did not diminish after this. Later, Cantor reaped substantial recognition for his work, including honorary membership in the London Mathematical Society, membership in the Society of Sciences at Göttingen (one of the great centers of nineteenth century mathematics), and a medal from the Royal Society of London. Unfortunately, all this attention did not prevent him from dying in a mental institution at the age of seventy-two.

While the great body of Cantor's work has been accepted by twentieth century mathematicians, problems persist. First, there are still intuitionists who reject the idea of a completed infinite. Any mathematics that requires an infinite process coming to some end, as the natural numbers do in reaching ω, is suspect. In addition, some of Cantor's work has been shown to be plagued by logical paradoxes. We know that for any two finite numbers, a and b, one of the following relationships must hold: $a > b$, or $b > a$, or $a = b$. Is this always true with transfinite numbers? Cantor proved that since $n < 2^n$, then the alephs he generated (i.e., $\aleph_0, \aleph_1, \ldots$) shared one of the above relationships. But, are there other transfinite numbers that are not one of the alephs defined by Cantor? In fact, is c one of Cantor's alephs?

The problem of whether any two transfinite cardinals shared one of the three relationships of greater, smaller, or equal, hinged on the well-ordering hypothesis.

> DEFINITION: A set is well ordered if for every subset, including the set itself, there exists a first element. The empty set, $\emptyset$, is considered well ordered.

If the set of all cardinal numbers is well ordered, then it is possible to prove that they can be compared to one another. Cantor believed that his transfinite cardinal numbers formed a well-ordered set but could not prove it. The well-ordering theorem, that every set can be well ordered, was finally proved by Ernst Zermelo (1871–1956) in 1904, and it was possible to prove that for every two cardinal numbers, one of our three relationships holds.

The well-ordering theorem did not answer all the difficulties with Cantor's theory of transfinite numbers. The natural numbers culminate in the first transfinite number ω. Did the transfinite numbers come to an end in Ω? To assume they did led to a contradiction, and Cantor recognized that the transfinite numbers never reached a limit as the natural numbers do. He stated this in the following theorem.[8]

THEOREM: The system Ω of all numbers is an absolutely infinite, inconsistent collection.

It appears extremely odd to talk of Ω as if it were a unity, when, in fact, by its very definition, it is inconsistent to do so. This notion is troubling to many who see it as a basic flaw in set theory.

Even though problems still exist for set theory and Cantor's infinite numbers, the theory of transfinite numbers has added an exciting and important field to modern mathematics. When we want to think of something way beyond ourselves, we now have truly enormous numbers to ponder.

CHAPTER 13

The Genius Calculators

We have reviewed how human beings struggled to discover all the kinds of numbers that now fill our mathematical world. While much of this discovery occurred because of a methodical plodding by societies of ordinary men and women, some advances have been due to the efforts of singular individuals. How is our understanding of numbers related to our intelligence as human beings, especially the intelligences of exceptional people? If there were no exceptionally talented individuals, what level of mathematics would we have achieved? Have the exceptionally bright really made particularly revolutionary contributions? Can exceptional people manipulate numbers in special ways, thus achieving insights into numbers denied to the rest of us? Our first effort will be to review some cases of individuals who seem to have had a calculating ability far beyond the average person.

Beginning in the early eighteenth century, records were kept of numerous individuals who could perform difficult and rapid mental calculations, many of them child prodigies. Some of these people grew up to become brilliant scientists and mathematicians, while others were retarded. This broad spectrum of brain power suggests that the ability to perform mental calculations is not a characteristic of general intelligence but is a specialized talent. Today we are seldom awestruck by the quick mental manipulation of large numbers. Possibly this is a reflection of a more sophisticated and jaded world, or it may be due to the rapid growth of computers, machines that tend to trivialize computations. Why should we practice multiplying six-digit numbers in our heads when our home computers can easily multiply fifteen- or twenty-digit numbers? It is my hope that we have turned away from the tedious grinding out of ponderous computations for their own sake in favor of asking more abstract and meaningful questions. During previous centuries, mental arithmetic was a frequent item on school curriculums. To exercise their memories, people often took the time to memorize long poems and songs. Perhaps mental dexterity was more valued in those less sophisticated years.

Even though we do not emphasize mental arithmetic now, we may want to review exceptional cases of mental prowess if only to gain greater understanding of how we, as human beings, add our own bias to the concept of numbers. Fortunately, two excellent resources are available for our investigation. For the exceptional calculator among the normal population we have *The Great Mental Calculators*, by Steven B. Smith, a professor of psychology at Brown University.[1] For a study of great calculators among the mentally impaired we have *Extraordinary People*, by Darold A. Treffert, M.D., director of a mental health center in Fond du Lac, Wisconsin.[2]

CALCULATORS AMONG THE NORMAL

Calculation with large numbers by normal or gifted people is generally characterized by a set of specific skills and circumstances. To begin with, most calculators were child prodigies and have, or had, an exceptionally developed memory. These people often grew up in isolated circumstances and, in their solitude, learned to love numbers. They frequently appear to be compelled to count and calculate. Often, when the special talents of prodigy-calculators are discovered, they or others make an effort to show their talents to the public. Hence, numerous calculators became stage performers, entertaining audiences in Europe and America.

The kinds of calculations performed fall into a narrow set of classes. First, multiplication is a dominant theme in their performances. Two large numbers may be multiplied together or a number may be raised to a power. These demonstrations usually involve numbers of three or four digits either squared or cubed, or a one- or two-digit number raised to higher powers. In addition, the extraction of roots from large numbers is also popular, although, surprisingly, the extraction of roots is less complicated than generally realized by the viewing public.

In addition, performing calculators may also demonstrate the ability to identify large prime numbers or to factor large composite numbers, although this skill is less frequently displayed than those involving multiplication and root extraction. Some calculators also demonstrate the ability to memorize large numbers or large collections of numbers, or to identify the cardinal number of a set of objects at a single glance.

Seldom do calculators demonstrate examples of addition or subtraction. Division is used for factoring, but is not demonstrated for its own sake. It appears that the multiplicative relationships between numbers are attractive to calculators and, therefore, easier to remember.

Just how great are these abilities? On the surface, they appear to be mind-boggling. For example, Jedediah Buxton (born in 1702 in Elmton, England) was reported to be completely illiterate and possibly retarded. Yet, he squared a thirty-nine-digit number in his head. He took two and a half months to complete the task, but when we realize he could not read or write numbers, it is a spectacular feat. Buxton performed difficult mental calculations for neighbors in exchange for free beers. He could remember the number of beers thus obtained from each patronized pub since the age of twelve.

Johann Martin Zacharias Dase was born in Hamburg, Germany, in 1824. During a test of his abilities, he multiplied two eight-digit numbers together to get the correct sixteen-digit answer in fifty-four seconds.

Although mental calculation is rare among women, there have been a few great female calculators. In 1980 Shakuntala Devi of India demonstrated her abilities by correctly multiplying two thirteen-digit numbers together to get a twenty-six-digit result in only twenty-eight seconds. The two numbers were randomly selected by computer. As reported by Steven Smith, "Such a time is so far superior to anything previously reported that it can only be described as unbelievable."[3]

Extracting roots from large numbers is a popular performance by calculators. There are two reasons for this: First, it makes for a good demonstration, and second, it is less difficult than one might imagine. For example, the cube root of an eight-digit number will be only a two-digit number. Hence, the extraction consists of determining the root's ending and beginning digits. As the number of digits increase in the power, the digits of the root become increasingly hard to compute. The numbers used for root extraction are always perfect powers so the roots will be whole numbers. For odd powers, the last (right) digit of the root can be uniquely determined by the last digit of the power. A trivial example of this is the fact that all numbers ending in five have five as a divisor. Hence, a power ending in five must have a root ending with a five.

Wim Klein was born in 1912 in Amsterdam. During his professional life, he performed all over Europe, finally going to work as a numerical analyst for CERN (European Organization for Nuclear Research). In 1974 he extracted the twenty-third root of a two hundred digit number in eighteen minutes and seven seconds. But this was not his most impressive feat for he later extracted the seventy-third root of a five hundred digit number in two minutes and nine seconds.[4]

An outstanding memory is also demonstrated by the calculators. Hans Eberstark (born in Vienna in 1929 but raised in Shanghai) has memorized the first eleven thousand digits of the decimal expansion of π. Results of psychological tests[5] show that most people have a rather limited memory for numbers. The average number of digits that can be recalled from a single learning session is seven if the numbers are pronounced in a monotone voice, and up to nine if pronounced rhythmically. This can go up to twelve numbers if the test numbers are pronounced both rhythmically and in pairs. The Italian calculator Jacques Inaudi (1867–1950) was tested by the famous nineteenth century psychologist, Alfred Binet, who determined that Inaudi could remember forty-two digits at a single hearing. This is three to six times what normal individuals achieve—impressive, but not unbelievable. Salo Finkelstein (born 1896 or 1897 in Lutz, Poland) could repeat from twenty to twenty-eight-digit numbers after only a one-second exposure. He memorized a thirty-nine-digit number in only four seconds.

Memory obviously plays a great role in calculating with large numbers. Wim Klein, by constant exposure, has memorized the multiplication table up to one hundred by one hundred (rather than the twelve by twelve of schoolchildren), the squares of all integers up to one thousand, and all the 1,229 prime numbers that occur through ten thousand. The ability to count rapidly is also found among some calculator's talents. Johann Dase could count in a single glance the number of peas thrown onto a table. He sometimes used his talent to total the number of books in a new acquaintance's library.

PRIMES AND COMPOSITES

The final talent we will consider is factorization of composite numbers and the identification of primes. In solving certain problems, some lightning calculators take a large number and break it down into its prime factors. For example, the number 30 can be "broken down" or factored into 2·3·5. Why are the numbers 2, 3, and 5 called prime numbers?

> DEFINITION: A *prime number* is a natural number that can only be evenly divided by itself and the number 1.

Other, nonprime numbers can be evenly divided by more numbers than 1 and themselves. For example, 4 can be evenly divided by 1, by 4, and by 2. The number 6 can be divided by 1, 6, 2, and 3. Such numbers are composite numbers.

> DEFINITION: A *composite number* is a natural number that can be evenly divided by more numbers than 1 and itself.

From the above definitions, it is easy to see that all natural numbers larger than one divide themselves into two classes: prime numbers and composite numbers. We cannot factor a prime number into smaller numbers, but all composite numbers can be expressed as a collection of prime numbers multiplied together. This leads to one of the great cornerstones of mathematics, which says that the way a number factors is unique; that is, each number has a unique set of prime factors.

> FUNDAMENTAL THEOREM OF ARITHMETIC: Every natural number greater than 1 can be expressed as a product of prime numbers in one and only one way.

Some prodigy-calculators, in order to multiply two large numbers, factor the numbers before manipulating the individual primes to arrive at their answer. Yet, mathematicians recognize a deeper question dealing with prime numbers than simply using them in calculations. An infinite number of natural numbers are prime, but the prime numbers thin out as the natural numbers get larger. The problem of quickly identifying which large numbers are prime and which are composite has intrigued mathematicians for centuries. What about the average person's ability to recognize prime numbers? The novice, after first learning what a prime is, has no difficulty in identifying 7 as a prime and 9 as a composite. Anyone practiced with numbers can recognize at once that 51 is composite (3·17) while 53 is a prime. Professional mathematicians can frequently identify three-digit and four-digit numbers as primes or composites.

Factoring large numbers and recognizing large numbers as primes is of interest because there is no known algorithm, or precise sequence of operations, to distinguish quickly a large composite from a large prime. Yet, some calculators seem to have a knack for it. Zerah Colburn (1804–1840) was an American calculating prodigy born in Vermont. He could take any six- or seven-digit number and tell if it was composite or prime. If composite, he could list its factors in a few seconds. His algorithm was subconscious, but upon careful reflection he determined that he had memorized an extensive set of number endings to identify factors. For example, only certain digits in the two rightmost positions of factors will result in the corresponding ending of the composite. This is similar to the technique used for extracting roots.

For example, Colburn's list for potential ending digits of factors to the composite 2983 consists of twenty pairs of numbers. That is, 83 can only be formed by twenty different two-digit pairs. By inspecting these pairs it is possible to discover the factors of 2983. The first pair of ending digits is 01 and 83. Since 1 is not a factor (by virtue of being a factor of every number), possible factors could be 101 and 83. But their product is a number greater that 8,000. Since appending any other digits to 01 and 83 only produces even larger numbers, we can eliminate this first pair. The second pair is 03 and 61. We can see at once that 3 is not a factor of 2983 (by adding the digits and discovering they are not divisible by 3) so the next possibility is 103×61. However, we eliminate this combination on the same grounds as the first. Finally, when we get to the eighth pair, 19 and 57, we find our factors. The number 19 cannot be made a 119 or the resulting product will be too large. But realizing that 20×150 is 3,000, we see at once that 19×157 is our 2983. Of course, for us to use such a factoring table is slow and clumsy, but in the hands of a prodigy-calculator, especially one who has internalized the whole process, the calculation is easy and appears almost simultaneous.

The two-digit table memorized by Colburn and others is not sufficient, and additional help is needed for factorization. George Parker Bidder (1806–1878) was a calculating prodigy from Moreton, England. He stumbled onto a factoring algorithm already known to mathematicians. Only odd numbers are of interest in factoring problems since any even number is divisible by two and can be factored and reduced to a smaller composite. An odd composite can be written as the product of the following: $(a + b)(a - b)$ which, when multiplied gives $a^2 - b^2$. Therefore, every odd composite number can be written as the difference between two square numbers, a and b. Bidder used this fact to test for factors. If x is the number we are after, then $x = a^2 - b^2$ or $a^2 - x = b^2$. Therefore, we simply look for an "a" whose square differs from x by a perfect square.

Let's try the method on the number 5251, which has the two factors of 59 and 89. We begin with the smallest perfect square above 5251, which is the square of 73. Of course, it would help to have memorized the squares of all numbers from 1 to 100! Now we square 73 to get 5329 and subtract 5251. The result is 78—not a perfect square. Now we increase 73 by 1 to 74 and square it to get 5476. Subtracting 5251 we get 225, which is the square of 15. Therefore, we can write $5251 = (74 + 15) \cdot (74 - 15) = 89 \cdot 59$—just the two factors we were looking for.

In practice we do not have to keep adding 1 and squaring to our test "a." Notice that $(a + 1)^2 = a^2 + 2a + 1$. Hence, to increase the square of a by 1, all we need do is add $2a + 1$ to a^2.

Using the two techniques of number endings and difference of squares, calculators can quickly factor numbers up to one million and sometimes to ten million.

THE CASE OF JOHN AND MICHAEL

A special class of prodigy-calculators, called idiots savants, have shown the ability to recognize exceedingly large numbers as prime or composite. Idiots savants are individuals who have a significantly low general intelligence as measured by standard I.Q. tests, yet possess a remarkable talent in either music, art, or calculating. Savants are frequently those who are severely autistic or have suffered left-side brain damage during their childhood. The exceptional musical or number recognition ability is generally combined with an outstanding memory. Some savants can remember the weather for every day of their lives. The calculating talent usually manifests itself as calendar calculating where the savant can give the proper day of the week if you give him or her the year, month, and day for some date in the future or past. Another incredible and rare talent for savants is the ability to identify prime numbers. This is the case with John and Michael.

Doctor Oliver Sacks is a professor of clinical neurology at the Albert Einstein College of Medicine, New York. He first met the savant twins, John and Michael, in 1966 at a state hospital. The twins at that time were twenty-six years old, having been institutionalized since the age of seven. They were both autistic, psychotic, and retarded. Yet, Doctor Sacks discovered the twins possessed a strange talent as he watched their normal activity.

One day, as Doctor Sacks watched the twins, he noticed they were engrossed in a strange conversation, each of them speaking in turn, followed by a period of silence. The boys both wore smiles on their faces, as if they were enjoying some private joke. He quietly approached, and sat down to listen. First John would say a six-digit number. Michael would hear the number and then concentrate. In a moment, he would answer with his own six-digit number. It was then John's turn to contemplate before responding. From the looks of pleasure on the boys' faces, Doctor Sacks knew they were enjoying their little game.

What were these numbers? Doctor Sacks wrote some of them down and took them home for study. Conferring with various mathematics books he discovered that all the numbers were primes! A six-digit number is between one hundred thousand and one million. How could these two young men, both retarded, know of, or recognize, such large primes? But the story does not stop here, for Doctor Sacks returned to the hospital the next day with his own book of prime numbers.

This time he sat down with the twins so he could join in their game. At first they hesitated to play in front of the doctor, but soon resumed. After silently watching for a time, Doctor Sacks offered his own eight-digit prime, obtained from his book of primes. They both stared in surprise at the newcomer. After a pause lasting for a half-minute or more both twins suddenly smiled and nodded their heads. John concentrated for a long time and finally gave a nine-digit prime. After a pause to think, Michael gave a second nine-digit prime. Doctor Sacks checked in his book of prime numbers and offered a ten-digit prime.

Nine- and ten-digit primes! A nine-digit number is between one hundred million and one billion. A ten-digit number is between one billion and ten billion. The game continued, but soon the boys were beyond Doctor Sack's ability to follow.

> . . . then John, after a prodigious internal contemplation, brought out a twelve-figure number. I had no way of checking this, and could not respond, because my own book did not go beyond ten-figure primes. But Michael was up to it . . . and an hour later the twins were swapping twenty-figure primes, at least I assume this was so, for I had no way of checking it.[6]

Twenty-digit primes? Could John and Michael really identify such huge primes? How can anyone, let alone someone with a low I.Q., identify a twenty-digit prime number? I have a 386-25 MHz computer. Even with double-digit precision operations in BASIC, I cannot test numbers with my computer beyond the ten trillion range (thirteen digits) for primeness (unless I am willing to turn my computer on and leave it on for days at a time). How did these two savants even remember a twenty-digit number, let alone test it for primeness? If the reports are true, then these individuals achieved something truly remarkable—not remarkable just because they are retarded, but remarkable when compared with anybody's talents.

Authorities who have studied the histories of John and Michael agree that the two young men could not, and did not, find primes by testing large numbers in the classical way. That is, they did not identify primes by mentally dividing large numbers by smaller prime numbers. They could not carry out such mental computations because they could not even add small numbers with any success. Besides, the number of mental computations necessary to test even a six-digit prime would be beyond them. It is also unlikely that they consciously used any method such as that employed by George Bidder, who represented odd composites as the difference between squares, since this method uses addition, subtraction, and multiplication extensively.

So, how did John and Michael do it? We simply do not know. One intriguing idea is that prime numbers possess a quality that can be somehow visualized to identify them as prime. Maybe the twins could visualize numbers and see this quality. Possibly, we ordinary human beings can learn to "see" this quality. Let us hope that someday we will figure out how John and Michael performed their astounding feats. Unfortunately, we will have to rely on other savants or other groundbreaking research to crack the secret. In 1977 the twins were separated by their guardians in order to promote their socialization. Although they both now demonstrate an improved ability to function with supervision in the environments of halfway houses, neither of them now performs any meaningful calendar calculating or prime identification.[7]

WHAT DOES IT ALL MEAN?

Some would claim that there is no difference between a prodigy-calculator and normal folks except hours of practice, a love of numbers, and a good memory. W. W. Rouse Ball (1850–1925)—historian, mathematics lecturer and former fellow of Trinity College, Cambridge—defends this position in his chapter on "Calculating Prodigies."

> The performances were so remarkable that some observers held that these prodigies possessed powers differing in kind from those of their contemporaries. For such a view there is no foundation. Any lad with an excellent memory and a natural turn for arithmetic can, if he continuously gives his undivided attention to the consideration of numbers, and indulges in constant practice, attain great proficiency in mental arithmetic, and of course the performances of those that are specially gifted are exceptionally astonishing.[8]

Steven Smith has much the same belief: He claims that the mental faculties used are the same as those used for language. Hence, any individual with a sufficient desire can learn to be a calculator.

> But if you were motivated enough . . . and if you knew the multiplica-
> tion tables I could teach you how to extract cube roots in less than a
> week and it would take me a month or two to train you in the
> fundamentals to become a calculating prodigy.[9]

Are Ball and Smith correct in their assessment that *all* calculators are basically the same as normal people? Or, are there some calculators, such as John and Michael, who fall outside the normal range of human abilities? Regarding the normal or bright calculators, Smith is probably correct in his assessment. Given the proper motivation and situation, most people could become calculators. The skills demonstrated are fascinating, but seem to group around a set of skills involving exceptional memory (which can be acquired by training), a love of numbers, and plenty of time to play with number relationships. Relationships that are discovered at an early age are frequently incorporated into calculating algorithms that later become subconscious and almost instantaneous.

But when we come to the autistic and retarded individual, Smith's hypothesis is simply not enough. Dr. Darold Treffert suggests that a combination of factors combine to create a musical or calculating savant—the Savant syndrome. When we consider that idiots savants frequently cannot multiply two single-digit numbers, or even add and subtract, we see at once that playing with numbers and then internalizing the play is an insufficient explanation. Calculating savants generally have three number talents: (1) memory of numbers, which includes the memory of days and dates as in calendar calculating, (2) the ability to recognize primes and factors, (3) the ability to perform rapid or instantaneous counting. These are likely subconscious abilities, for the savants know perfectly well what they are doing, but they cannot describe how they do it.

Many savant calculators have diminished capacity of the left hemisphere of the brain. This can be caused by traumatic brain injury during pregnancy or during early childhood. For male children, it may even be the result of testosterone imbalance, which retards the growth of nerve connections in their left hemisphere. Treffert suggests that this early damage to the left brain results in a shifting of functions to the right brain. Near the time of birth billions of brain nerve cells fail to make connections to other cells and die off, and are reabsorbed by the brain. Evolution seems to have led to a production of excess nerve cells to ensure there will always be enough to make the necessary number of connections; the extra, unconnected cells are left to die. Perhaps when left-brain functions shift to the right brain, some of these excess cells do make connections to result in individuals with more powerful right hemispheres.

The dominant left brain in normal people is responsible for linear processes that are involved in speech, mathematics, reasoning, and planning. The right brain is dominant for instantaneous activities such as immediate visual apprehension. This division of functions between the left and right brain helps explain the differences between the normal individual and the calculating savant. The savant has trouble with linear processes (i.e., processes consisting of steps that occur one at a time), except for music, which, incidentally, is dominated by the right brain; and he or she is strong in problems requiring immediate awareness, such as instantaneous counting.

Therefore, for calculating savants we have more than just isolation, an interest in numbers, and practice to explain their abilities. They possess a brain that functions differently, one dominated by the right instead of the left hemisphere. Treffert also points out that their memories are different because of disruptions in subcortex areas. Their memories tend to be very intense, but narrow. This explains in part why so many savants appear to be obsessive in their thoughts. Yet, even this is not enough, for Treffert says:

> The talented savant can, with constant repetition and practice, create sufficient coding so that some unconscious algorithms develop (although he will have no understanding of them). However, in the prodigious savant, access to the rules of music or rules of mathematics, for example, is so extensive that some ancestral (inherited) memory must exist to account for that access. Such memory, in these individuals, is inherited separately from general intelligence.[10]

Treffert has introduced an added element: an inherited memory of mathematical or musical rules. Yet, Treffert does not mean by memory the recall of specific past events. He is using memory to refer to inherited neurological skills for performing mathematical and musical operations. These special skills, he claims, are different from our general intelligence, and, therefore, when the general intelligence of the savant is damaged, these specialized skills are amplified. Based on his research, Treffert claims that the calculating savant is significantly different from a normal individual. With a right-dominant brain, possibly including a more complex right hemisphere due to increased neuron connections, access to normal inherited skills for mathematical rules, plus the environmental factors of psychological isolation and an interest in numbers, the calculating savant reaches new depths of immediate number awareness.[11]

FUTURE CALCULATING

One may legitimately ask: Why all this bother about calculating? We can just turn on the computer and—bang—we get our answer. But calculating may be more important than grinding out an immediate answer to some large arithmetic problem. It may well be a foundation for learned mathematical insight and a deeper awareness of numbers.

Historically, a number of outstanding scientists and mathematicians have been prodigy-calculators, including the great mathematicians Leonhard Euler, Carl Gauss, and Srinivasa Ramanujan. John von Neumann, the father of modern computers and game theory (the mathematical theory of best choices), was also known for his calculating ability.

The talented calculators all have one characteristic in common: They love thinking about and playing with numbers. This desire for number play has been a critical element in numerous mathematical breakthroughs. With the early Pythagoreans we saw how their play with collections of pebbles suggested certain number relationships. Their number play led to some of the earliest mathematical theorems. On a more recent note, one of the greatest of all mathematicians, Carl Gauss, made a most important discovery in number theory in 1792 when he was only fifteen years old.[12] He suggested that the number of primes found in the natural numbers increases as a function of $n/(\log n)$. This function later became the Prime Number theorem. How did Gauss think of this function? He counted the number of prime numbers found in successive blocks of one thousand natural numbers. This distribution suggested the Prime Number theorem to him. Hence, Gauss, one of the world's most brilliant minds, made his discovery by diligently identifying and counting primes, routine calculating by most standards.

We will never know how many of our great discoveries in mathematics were made by someone willing to sit and patiently make mundane computations, but the number must be substantial. This same patience and diligence are reflected in the sciences where researchers have spent years collecting and analyzing data for one meaningful breakthrough. Therefore, simple calculation is essential to mathematics. Some critics of the twentieth century may despair that computers have stolen our desire to calculate, and we are, therefore, headed for a new dark age of mathematical blindness since future generations will soon lose the skill for arithmetic manipulations. I maintain the reverse will happen. Computers are not denying us the skills to manipulate numbers. Rather, they are an asset because they act as a calculating extension to our own brains. Now, each of us can be calculating masters with our home computers. The acts of multiplying or factoring are not in themselves as important as the new insights into numbers that they provide. Questions about numbers that the average person may never have considered before, because they entailed complex calculations, can now be answered with a few minutes or hours of programming. All that is required is an interest in numbers.

What Does It All Mean?

DEFINING THE QUESTION

We have come a long way in our quest for numbers. Beginning with the natural numbers, we have investigated their origins and expanded our ideas to include fractions, irrational, complex, and transfinite numbers. We have even speculated on how other species count. Now it is time to ask: What are numbers in the most general sense? This is really not a mathematical question but a philosophical one, and as such, we must be careful not to get bogged down in a quagmire of endless philosophical debate.

Most concepts in mathematics are easy to understand since they are based on simple definitions and specific relationships that we can quickly grasp. The philosophy of mathematics is much more difficult because it forces us to step back from the proving and calculating of mathematics and question our intuitive concepts by asking deeper kinds of questions. Our primary concern about numbers is this: In what way do numbers exist, and what does this imply about the universe we live in? If *ontology* is the study of being or existence, then we want to understand the ontology of numbers. Here we come to the cornerstone of the entire issue: If numbers exist, how do they exist, and do they exist independently of human beings? We do not mean, of course, that numbers exist in the Pythagorean sense, that is, as physical objects floating around in space–time. We will never find the number three resting under a rock or hiding behind a tree. What we mean is: Do mathematical objects (whatever these objects really are) exist in relationships independent of our thinking of these objects and relationships? In other words, do the truths of mathematics exist as truths about the universe in which we live, regardless of our own existence?

You can see at once that we have opened the door to the sticky concept of existence. Various philosophers have struggled for twenty-five centuries to make some kind of sense out of the idea of existence, and whole philosophical movements have been founded on specific interpretations of ontology. We must not be tempted to rush forward and give a glib answer to all the questions implied by ontology, for to do so would be to enter a mine field of vacuous speculation. In these short considerations we cannot hope to settle all the philosophical questions regarding the being of numbers. Perhaps the best we can do is to outline the questions regarding numbers in such a manner that they lead to clearer thinking. In this connection, we shall agree that we are not going to come up with any final answer regarding numbers, yet we will not be so timid that we fail to speculate freely about their ontological status either.

As a background to our speculations, we will define minimum and maximum ontological positions regarding numbers, and hope that the final answer will be trapped between these two extremes. First the minimal position: Numbers exist only in the minds of human beings. If human beings did not exist, then numbers would not exist. Or, to be more precise, if some thinking entity did not exist, then numbers would not exist. Numbers, along with all other mathematical truths, are simply thoughts we have, and as such, their existence as thoughts is not related to the existence of anything independent of our minds. The physical universe is "number free," and human beings superimpose numbers onto this numberless universe. This minimum position in mathematics is called *constructivism*, and in philosophy, *realism*. It emphasizes that we mentally construct numbers—they are not part of the nonhuman universe.

And now for the maximum position: Numbers are objects of thought. They are not the thoughts themselves, but rather the objects that the thoughts consider. All objects of thought (including numbers) exist independently of individual acts of thinking. These thought-objects are external to the mind and possess a perfect existence superior to the changing physical universe. Hence, numbers and other thought-objects form an ideal universe of truth that is reflected in the inferior, ever changing, material universe. In fact, the physical universe is nothing but an illusion, and what has real existence is the world of ideas. This maximum position is *classical idealism* or *Platonism*.

Idealism and realism have been only briefly defined, and no attempt has been made to justify either. Before continuing, it will prove helpful to introduce some additional terms to keep our discussion from degenerating, and we use the terminology consistent with the great twentieth century philosopher and mathematician, Bertrand Russell.[1]

- *Sense-data*: the sensory input from the world around us, that is, sounds, colors, odors, texture, and so on.
- *Sensation*: our consciousness of sense-data; that is, our immediate awareness of experiences of red, hard, sweet, and so on.
- *Physical objects*: these are the hypothetical objects we believe exist in a physical world and we assume they are the cause of our sense-data. An apple is a physical object we deduce is before us because we have a sensation of specific sense-data of round, red, and hard.
- *Idea*: an idea is an object of thought. Ideas do not come from sense-data, nor do we have sensations of them. We think of them internally.
- *Thought*: this is a specific awareness of an idea.
- *Universal*: objects existing in a world of universals that we suppose are the cause of the specific ideas in our thoughts. These universals include numbers, mathematical relations such as π, and possibly other general attributes, such as virtue, freedom, greenness, loudness, and so on.

We now display these definitions in Figure 52, which shows our consciousness within a large box. Inside our consciousness we directly experience sensations of sense-data and then assume these sense-data imply the existence of physical objects outside the box. In a corresponding manner, we have thoughts of ideas. Do universals exist outside the box of consciousness that correspond to our internal ideas? Of course, we are just as justified in asking: Do physical objects exist that correspond to

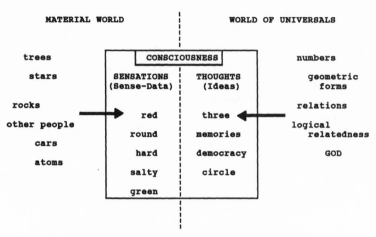

FIGURE 52. Symbolic representation of consciousness.

our internal sense-data? Philosophers maintain we can never really be certain of what is outside the box of consciousness, for everything we experience, the sense-data and thoughts, are experienced inside the box. For this reason we have used a heavy solid line to separate both the world of universals and the material world from our conscious world. The division between sense-data and thoughts is a broken line to indicate our uncertainty as to the exact distinction between sense-data and thoughts. The same can be said of the distinction between the two external worlds. Are they really separated, and if so, how? Figure 52 is, of course, a simplified schema, and leaves many questions unanswered. For example, are memories thoughts or internal sense-data? What are emotions?

Most of us hold a view of the world known as *naive realism*. This view says that the material objects in the universe exist exactly as we perceive them. The tree I see outside my back window really is tall and cylindrical, with brown bark and green leaves. However, when we take a closer look at naive realism, problems at once confront us, and we are led to the conclusion that material objects are not exactly as we perceive them. This argument was given its classic interpretation by the eighteenth century philosopher, Bishop Berkeley (1685–1753), who in his work, *Three Dialogues Between Hylas and Philonous*, expounds on the notion that those attributes we commonly assign to material objects are really attributes specific to human sensory organs. From this, Berkeley concluded the world was not material but mental in nature.[2]

The German idealist, Immanuel Kant, gave us terminology to help describe this discrepancy between the sense-data and the implied objects that supposedly cause them. The sense-data he called the *phenomena* and the objects generating them are the *noumena*. He argued that since we can only truly know the phenomena, the noumena will always remain a mystery to us.

Under the strictest interpretation of the constructivist view, we would say that everything outside the box in Figure 52 is a physical object. When the individual dies, the box collapses and leaves a world of stars, planets, and rocks. This is simple materialism. With classical idealism or Platonism we say that everything outside the box of consciousness is a universal, and these universals generate both the sense-data of our sensations and the ideas of our thoughts.

The problem in determining which world view to accept is that believing in either one results in philosophical difficulties. We can opt for the middle road and accept that both physical objects and universals exist—a view known as *dualism*. However, such a maneuver does not remove the difficulties. Yet, most people would be more comfortable with this middle course. Most of us reject as a silly notion that the physical universe does not exist. On the other hand, we become uncomfortable with only a materialistic view of nature. Therefore, it seems safe and comfortable for us to accept the physical world while also believing in a second world of universals. We frequently think of this world of universals as a spiritual realm that contains not only the universals but also spiritual entities such as departed souls and even God.

Yet, the situation may be even more complex. Look at Figure 53. Here we have shown a kind of filter between the outside of the box and the inside. This reflects the idea that our conceptualization of the world, both the world of universals and the world of physical objects, is dependent on our own architecture as human beings. That is, we see things and think of things the way we do because of how we are constituted as beings (either

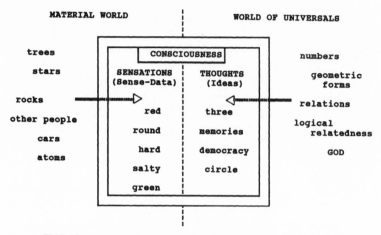

FIGURE 53. Symbolic representation of consciousness with a filter.

physical or idealized beings). This is known as the egocentric predicament. In a biological sense, we understand the egocentric predicament by
realizing that the very structure of our brains, with its billions of
interconnected neurons, helps determine the format of how we perceive
reality. How much does the architecture of our consciousness determine
how we see things? Do we believe that physical objects exist in specific
ways primarily because we have this architecture? Are numbers the
objects of our thoughts because of this same architecture?

If we wish to be conservative about the ontology of numbers, we
may adopt a view that we can only be certain of the sense-data and the
internal ideas of our thoughts, for only these things come to us in
immediate awareness. No one argues that I see green when, in fact, I am
experiencing green. The sense-data of green is a given. All else outside
the consciousness (if indeed it makes sense to talk of inside and outside
at all) has only a second kind of existence; an existence dependent upon
our immediate awareness. This is not to say that our immediate awareness
(sense-data and thoughts) causes these outside objects to exist, but that
we infer knowledge of the outside objects from sense-data and thoughts.

Why then should we believe in either physical objects or numbers

(numbers as classic representations of universals)? The reason for believing in physical objects is obvious: If we do, then we make judgments regarding actions that allow us to function successfully. If we try to ignore physical objects, then our ability to survive is impaired. Assuming the objective reality of the physical universe seems to work well in a world view that enhances our survivability.

What about numbers? Is there any reason to believe that numbers exist outside of our individual consciousness? Here we come back once again to the cornerstone of our inquiry.

HISTORICAL PROGRESSION

It is romantic to suppose that human beings have considered such questions as long as we have possessed brains powerful enough to do so. We can imagine individuals sitting under a shady tree tens of thousands of years ago arguing about whether the tree was really there or not. However, the first evidence that such philosophizing occurred comes from the ancient Greeks who had the foresight to record their thoughts so that one day we might read them. We have already seen that the Pythagoreans believed that physical objects were constructed directly from numbers. Hence, the physical objects outside of their (and our) consciousness had an atomic structure based on physical numbers. This meant that numbers occupied not only the world of universals but also the physical world. Plato believed in a world of ideas, containing perfect, eternal universals that somehow mix with space to give us the physical world.

Since the time of Plato, philosophers have joined different camps, some leaning toward realism while others argued for idealism. In both the western and eastern cultures, religion generally supported views favoring idealism. Combining views from both Plato and Aristotle, the Christian theologians of the Dark Ages claimed not only that God created the physical universe but that His infinite intelligence maintains the existence of all universals by constantly thinking about them. Hence, the idea of a circle would not cease if all human beings died, for the universal of circleness would remain in God's mind.

Even into the twentieth century we see this close relation between the idea of an omnipotent God and universal ideas. Cantor proved that there can be no largest transfinite number. All the transfinite numbers are less than the infinite and indescribable nature of God. God is the Absolute Infinity, with the lesser infinities of transfinite numbers between God and the finite world. Most mathematicians of the nineteenth and early twentieth centuries were idealists believing in the objective nature of mathematical objects, including numbers. What about now? How do mathematicians view the numbers that are at the center of mathematics?

If we stopped the average man or woman on the street and asked, "Do numbers exist independently of human beings?" we would probably receive a negative answer. If we rephrased the question as, "Does the ratio of the circle's diameter to its circumference exist independently of human beings?" we would probably get a mixed response. If we asked pure mathematicians both questions, most would answer, "Yes, they exist as truths prior to, or independent of, human beings." Hence, while the average man and woman on the street tend more toward materialism, mathematicians, and especially pure mathematicians, tend toward idealism. Why?

WHAT DO WE MEAN BY EXISTENCE?

If we want to understand whether numbers exist, then we should review what this property of existence is. When we say something exists, just what do we mean? Let's consider physical objects. Why do we believe material objects exist outside of our consciousness? First, there is the consistency of our sense-data. Whenever I walk into my living room, there is the couch. If I turn my back to it, and then turn around— there it is, again. The experience is consistent. If the couch were there one moment, but suddenly evaporated the next moment while I was still looking at the same location, I would begin to question whether the couch really existed or not. It is the same with other material objects in our world. They are there for us. If they disappear, we expect some meaningful explanation. They do not pop in and out of our world at random. Hence, material objects are represented by consistent sense-data.

Do mathematical objects have this quality? Mathematical objects do not occupy a specific place and time, so we do not expect to find them among our sense-data (although specific sense-data can suggest the universals). Yet, mathematical objects are consistent. The natural numbers stay properly ordered in their sequence. Their various relations stay fixed. The ratio of the diameter of a circle to its circumference is one number that does not change. (This assumes that an evil demon does not exist who constantly changes the value of π and also changes my memory of what it is.) Therefore, the truths of mathematics do exhibit consistency, just as do the sense-data of material objects.

A second feature of existence is independence. Objects in the physical world do not appear to rely on our thinking about them. This is dramatically illustrated when we venture into unknown territory. We walk along a path in the forest, excited by the prospect of being surprised by what is around the next bend. We do not already know what objects are waiting for us because these objects do not owe their existence to us. Hence, we are led to believe that if we do not walk around the bend, those out-of-sight objects will still exist.

Surprisingly, mathematics has this same sense of discovery, and it is the experience of constant discovery that leads many mathematicians to believe that the mathematical truths they pursue have an existence prior to their discovery and will continue to exist afterward. For example, if π is an invention of my own thinking, then why do I not know the millionth digit of its decimal expansion? To know what this digit is, I must carry out a number of mathematical operations and "discover" the appropriate digit. Of course, the constructivist would say that the millionth digit of the expansion of π does not exist until someone computes it. Before such a time, the digit is only a potential digit.

With this same sense of discovery, a Pythagorean of long ago discovered the diagonal of a square is incommensurable to its side. He or she did not define it so—it was discovered to be so. Even though mathematicians try to be critical of mathematical existential claims, they have a difficult time because of this psychological feeling that they are discovering and not inventing. While it is true that pure mathematicians invent new mathematical systems, they do it by inventing axioms, and then discovering the theorems implied by these axioms.

Is there an example of pure invention that we can contrast with the "discovered" world of mathematics? Consider our English language, which is a purely human invention. We treat dictionaries as collections of definitions and not tomes of truth. If suddenly we were to encounter a higher intelligence visiting us from the center of our galaxy, we would be shocked if it spoke English. This is because we know we invented English, and would be surprised to believe anyone else would independently invent the identical language.

But the situation is not as certain when it comes to mathematics. Would we be surprised if this other intelligence had no concept of number? Do we believe that numbers are so universal that all highly intelligent beings should be aware of them? We have even assumed in our science fiction that the perfect method of opening communications with extraterrestrial beings is through the "universal" language of mathematics. But if this is so, then the truths of mathematics do not rely solely on human invention.

However, this does not settle the case in favor of idealism. The fact is, we have not encountered beings from other worlds. Maybe if we did, we would be surprised that they had no art comparable to mathematics. Or perhaps only those beings that invent a mathematics similar to ours can evolve a science sufficient to travel through space and visit us! We have talked about the large brains of whales and dolphins. Can the blue whale, with its fifteen-pound brain, count the fish it swims past? Perhaps these ocean beings have a complex and intelligent culture that is devoid of any conceptualization of mathematical objects. We may think that mathematical truths are universal, and, hence, discoverable by all sufficiently intelligent beings, while in reality they may be only a reflection of how the human mind works.

The third property of existence to consider is the quality of the content of sense-data. We recognize that sense-data are not thoughts. When I see green, it has a qualitative richness that my thoughts about green lack. Our sense-data are strong and potent in a way our thoughts are not. From this fact alone we might infer that the objects behind sense-data truly exist, while the objects behind our thoughts do not. Yet, this is not conclusive evidence against idealism. Our mathematical ideas contain a sharpness that is lacking in sense-data. My idea of a circle (as the locus of points all equidistant from a single point in a Euclidean plane) is perfect and exact in a way that no sense-data can be. When I think of my circle, I may internally visualize a round set of dots, but these act only as an aid. They do not represent my idea of a circle, for my circle is a universal and not locked into space or time.

An amusing question is to ask where our thoughts of circles go when we are not thinking of them. When we turn our backs on the couch, the sense-data of the couch stops, but we believe the material couch continues to exist, ready to provide more sense-data as soon as we turn around. Where does the idea of a circle go when we are not thinking of it? Do we reinvent or rediscover it each time we bring it to consciousness? We suppose that the idea of a circle is stored in our memory, to be recalled when needed. If the idea of a circle is stored in memory, then there must be a specific collection of neurons in our brains that are configured to retain this idea. When the neurons fire, the idea of a circle comes to our consciousness. Hence, the concept of a circle has a biological analog in the material world. The analog must be exact, for when we think of "circle" we do not imagine a round thing that approximates a circle and then, from this memory, think of exact circleness. We actually have the memory of the precise idea or definition of circleness. If our idea of circle has an exact biological analog in the physical world, then all ideas have an exact analog in the physical world.

A modern example of dualism is presented by Bertrand Russell in his short book, *The Problems of Philosophy*. To establish his theory, he first demonstrates that universals exist, and speaks of them as being relations: either relations between physical objects or relations between other universals. He then says that the next problem is to decide if these relations exist independently of human minds—precisely our dilemma. He decided that they do, in fact, have such existence.

> We may therefore now assume it to be true that nothing mental is presupposed in the fact that Edinburgh is north of London. But this fact involves the relation "north of," which is a universal; and it would be impossible for the whole fact to involve nothing mental if the relation "north of," which is a constituent part of the fact, did involve anything mental. . . . Hence the relation "north of" is radically different from such things [physical objects]. It is neither in space nor in time, neither material nor mental; yet it is something.[3]

Russell concludes that since the universe is full of things, these things stand in relationships to each other. These relationships are not physical objects, nor mental 'thoughts', and therefore have an existence somehow independent of both. They exist in the world of universals. But Russell is too quick to draw his conclusion, and his argument lacks the kind of rigor that is convincing. A problem with his London–Edinburgh example is that he forgets that a mental activity precedes the relation of "north of" in this case. This activity is the definition of London and Edinburgh as physically distinct. Once we have defined them as such, we can then apply the relation of "north of." The problem here is in believing that the universe is full of distinct and separate things. The universe may, in fact, be a continuous manifold that only separates into distinct objects when some intelligence differentiates this manifold. Hence, the relations are dependent on how the differentiation takes place, and therefore exist only after such activity and not before.

Jerry King appears to be a modern constructivist. In his excellent book, *The Art of Mathematics*, he says:

> Mathematicians *do* mathematics. And when they do mathematics they deal with objects they have created. These objects are abstractions and have no existence outside of the imagination of the mathematician. They are endowed by their creator, the mathematician, with certain properties. From these assigned properties, using the laws of logic and the rules of mathematics, the mathematician deduces other properties.[4]

Yet King does not want to abandon the idealist position completely, for he leaves a slight crack open in the idealist door. He speaks of the strange fact that mathematics, as invented in the minds of mathematicians, models the material world in such a powerful manner.

> The creationist viewpoint fails to account, however, for the recurring applicability of some of the purest and most abstract of mathematical structures. Riemannian geometry, for example, follows with mathematical certitude from a set of axioms exactly as does the familiar geometry of Euclid.[5]

Elsewhere, King introduces the idea that there is actually some kind of mystical principle working for the more elegant mathematical forms to relate to their own universality.

Furthermore, there seems to be at work a high and mystical aesthetic principle which produces a positive correlation between the elegance of the mathematical idea and its correctness and importance—or, as Hardy put it, a positive correlation between the beauty of the idea and its *seriousness*.[6]

Just how we are to understand this mystical principle is not stated, for the very meaning of mystical implies a truth known by immediate awareness rather than analysis. If this mystical aesthetic principle produces a correlation between elegant mathematical ideas and mathematical correctness and importance, did mathematicians then invent this principle, or was it something existing prior to invention? Here we seem to be introducing a relation between objects, and then speaking as if this relation existed independently of human beings.

We may attempt to state the constructivist view, and hope to give it the credit it deserves. The universe is a continuous manifold that has no predefined divisions. In a sense, it is seamless. When we, as intelligent beings, experience the universe, we differentiate it into separate "things" based on our biological construction. Once we have done this, we define relationships between our differentiated objects. Hence, our defined relations (universals) are specific to us. Another intelligence may differentiate the universe differently, and hence come up with different relations, and thus different universals. Therefore, universals do not exist except in the minds of the specific intelligences defining them.

The above seems to be a good statement of the constructivist position. Why am I not completely convinced? Maybe you are, but I am still caught in a sea of uncertainty. If we assume that an intelligent species must first differentiate the manifold of the universe to have any knowledge of it (rather than apprehending the universe instantly as one organic whole—a very nice image), we are committed then to the idea that differentiation involves recognizing separate objects. Given that these particular objects are unique to us, as human beings, we still can group them into collections or sets. Once we have sets, we can ask for the manyness of the sets—numbers! It does not matter what objects we are talking about, only that we have separated the universe into these objects. Hence, an alien intelligence living in another galaxy may divide its world up in a completely different manner; and if it does any differentiation at all, it has sets and numbers, whether it recognizes them as such or not. Hence, it seems we cannot escape from the primary notion of sets and numbers. They seem to exist as a basic principle in the universe for any discussion of the universe as something containing objects. This is why I am not completely convinced of the constructivist argument. Yet, the opposite extreme of classical idealism makes me just as uncomfortable. Perhaps the safe road is dualism—accepting both the physical universe and universals.

The view that numbers are primal in the differentiation of the universe is also held by W. H. Werkmeister in his work, *A Philosophy of Science*.

> Whenever we open our eyes, our visual field discloses to us distinct forms and shapes and a variety of hues, shades, and tints; and various ideas are associated with what we see. That is to say, we are aware of distinguishable objects. This fact implies that we can segregate some specific object from all other objects. The segregated object is a self-identical "something" and, in this sense, it is a "one."[7]

THE EVOLUTIONARY PERSPECTIVE

We have not answered our original question: Do numbers exist independently of human beings? Nevertheless, we did not really think we would. The best we can say is that we have tried to define the question in a meaningful way. Sometimes it is helpful to look at a question in an evolutionary context. How has mathematics evolved as part of our culture, and what does this imply for the ontology of number?

Mathematics began when humans moved things about and noticed that there was a property that collections shared that was independent of the objects themselves—this property was the manyness of the collections. Hence, numbers were born (or discovered). At first, numbers were simply descriptives for finite collections of objects, such as a "pair" of gloves or a "brace" of oxen. The growth of mathematics depended on the process of abstraction. As numbers became more abstract, they became more universal—applicable to more and different collections. This abstraction has continued throughout our evolutionary history. In addition to abstraction, numbers became extended. First, there were only finite collections, and a finite number sequence. Then we realized the natural numbers were infinite—a concept not suggested by any sense-data (and an idea resisted vehemently for millennia).

But the extension continued. Numbers were extended to include fractions, then negative numbers. Irrational and transcendental numbers were added. Then complex, hypercomplex, and transfinite numbers were conceived. This dual process of abstraction and extension has marked the progression of our mathematical concepts. Can we assume we have reached the end? Of course, in every generation some people insist that the process is complete. How can anyone imagine going further? But, if we are at only one stage of our intellectual and cultural evolution, then we can assume that our mathematics and our notion of numbers are incomplete. The ideas regarding numbers will be pushed further into unknown realms.

This larger picture of human nature suggests a different solution regarding our puzzle over the ontological status of numbers. Let us assume that, in part, our notion of numbers is a reflection of who we are as human beings. That is, it is dependent upon our particular biological architecture. Now the question becomes, how much is purely human and how much is universal? We can see an analogous situation in geometry. At first, we assumed that the universe must be Euclidean in nature. Then we discovered non-Euclidean geometries. Does the universe have a Euclidean geometry or one of the non-Euclidean geometries? Einstein assumed a non-Euclidean geometry for his theory of general relativity. However, the geometry of the physical universe may not be a specific geometry as we have developed them. Space may be a more general manifold than any specific geometry, such that we can actually choose the geometry we wish.

In a similar way, the truths of the universe may be more general than the numbers and circles of our mathematics. If we were to encounter another society of highly intelligent beings, we might be confused and confounded by their specific mathematics. Behind their mathematics and our mathematics might lie a deeper truth, a kind of metamathematics that is generalized to a higher level. Here I am not using the word "metamathematics" as the logical study of the rules and principles governing the use of mathematical symbols, but I am suggesting a deeper, more generalized meaning. Both our mathematics and the alien mathematics would be specific instances of this metamathematics.

We have been trying to uncover the status of numbers, and in so doing we have relied on the logical processes of analysis. This is appropriate, since numbers are objects of thought. However, if we are to learn anything from our long history, we should learn that it is frequently dangerous to assume that the material world always operates according to the rules of logical inference. Ideas are related to each other in a logical manner, and sets of ideas can be consistent or contradictory. When we try to apply logic in a consistent way to every aspect of the physical universe we often are led down the path of error. To understand why, we must remember that our ideas of the physical world form a model, and we deduce logical conclusions within this model. Then we expect the physical world to agree with these conclusions. When our sense-data do not agree with the logical conclusions, we are frequently tempted to throw the sense-data away instead of trying to improve our model.

Parmenides and Zeno believed the world was unchanging, and motion was only an illusion because of their logical arguments. The sense-data of all the physical objects in motion were ignored. At one time it was reasonable to think the world was flat, for our model of reality allowed us to logically deduce that it was. If the earth were round, like a ball, it would fall into a great void. It was "logical" to believe that only ten heavenly bodies existed since ten was a sacred number, and all of the heavenly bodies had been accounted for. Therefore, it was justifiable to imprison Galileo Galilei for claiming he saw moons around Jupiter. Galileo's sense-data had to be in error, since the existence of more than ten heavenly bodies would be a contradiction.

If we can be sure of anything, it is that the universe is not compelled to act out the conclusions to our logical deductions. When we experience sense-data contradicting our models of reality, we must rethink our models. If the universe were perfectly rational, then the ratio of the diameter of a circle to its circumference would probably be three—clean, precise, and uncluttered—just a three. At least, if I had invented the universe, I probably would have picked π to be three. But I am glad I did not invent the universe. For me, the discovery of the true value of π is much more exciting and interesting than any number I could have dreamed up.

CHAPTER 15

Numbers

Past, Present, and Future

INTO THE TWENTIETH CENTURY

We ended our formal development of numbers with the work of Georg Cantor at the end of the nineteenth century on his transfinite numbers. This represents the last extension of the idea of number, but not the end to its story. In addition to the definition of irrational numbers by Dedekind and of transfinite numbers by Cantor, there were other great mathematical developments occurring during the last half of the twentieth century. One was the discovery of non-Euclidean geometries, and the second, the one we are concerned with here, was the attempt to establish all of mathematics on axioms.

The model for this effort was the work of Euclid, who used a set of axioms and postulates to deduce the theorems of classical geometry. Mathematicians hoped that a set of axioms would be found from which all the truths of arithmetic could be logically deduced. If this were possible, then all the various fields of mathematics could then be deduced from arithmetic, and the entire edifice of mathematics might be put, once and for all, on a sound foundation. The set of axioms must have two properties: They must be complete, that is, sufficient to give all the relevant properties of arithmetic, and they must be consistent. To be consistent, the theorems proved from the axioms must never contradict each other. If it were possible to deduce contradictory theorems, then how could anyone have faith in mathematics?

In 1889, the Italian mathematician Giuseppe Peano formulated a set of five axioms that could be used to deduce the theorems of arithmetic. These simple axioms were

1. One is a number.
2. If x is a number, the successor of x is also a number.
3. One is not the successor of any number.
4. If two numbers have equal successors, they are equal.
5. If a set of numbers contains the number one and it contains all the successors of its members, then the set contains all the numbers.[1]

In these axioms, Peano left undefined the notion of number and of successor. Yet, it is obvious that numbers, as used by Peano in his axioms, mean the natural numbers. From these simple axioms, and beginning with the first natural number one, he was able to define all natural numbers. Once we have the natural numbers, it is possible to deduce the rest of the numbers and their relationships. What is surprising is that such simple axioms are so powerful. They are not convoluted or obtuse, but embody the very notions basic to the natural numbers: that we begin with one and then move to each successive number in turn. The fifth axiom is the axiom of induction. It simply means that, if we can demonstrate a property for one plus all the following successors, it is then demonstrated for all natural numbers. These axioms, as used by Peano and other mathematicians, were written in symbolic form accompanied with strict rules of logical inference. Here we have stated the axioms in their basic English equivalents.

The process establishing an axiomatic foundation to mathematics was continued by Bertrand Russell and Alfred North Whitehead (1861–1947) with their joint effort to connect all axiomatic mathematics to formal logic in *Principia Mathematica*. If this could be accomplished, then all of mathematics could be traced back to basic axioms in logic. What could possibly stand in the way of mathematics now?

THE PARADOXES

Unfortunately, the program deriving all of mathematics from axioms did not progress the way Peano, Russell, Whitehead, and others had planned. Cracks appeared in the great edifice they were building. The first problem was the discovery of paradoxes that grew out of our intuitive notions of sets, and the idea of sets was at the heart of the new axiomatic mathematics, for sets were used to define the idea of numbers.

The first paradox we have already mentioned: It was Cantor's paradox, that there can be no largest transfinite number. Let Ω be the cardinal number of the all-inclusive set that contains all other sets as members. This gigantic set must contain not only itself, but also every possible subset of itself. Hence, the cardinal number of the set must be greater than Ω. It fact it must be at least 2^{Ω} which is a contradiction to the assertion that its cardinal number is Ω. This was handled, of course, by simply denying that such a largest set, or largest cardinal number, existed.

Russell discovered a paradox buried in naive set theory that was more difficult to explain away. We can construct sets whose members are sets themselves. If we let A be the set of all left-handed baseball batters, and B be the set of all drake ducks, then $\{A,B\}$ is the set consisting of two sets, namely the sets A and B. Russell formulated a set, S, whose members were themselves sets. He said that a set would be a member of S if it did not contain itself as a member. If $S = \{s_1, s_2, s_3, \ldots\}$ then every s_i is a set that does not contain itself as a member. Now, Russell asked, should we include S as a member of S? If S does not contain itself, then it should be a member. But if we make it a member, then it contains itself and should not be a member. But, if it is not a member, then it should be! Hence, the paradox. It is simply not possible to have a set such as S.

Russell's paradox demonstrated that our intuitive notions of sets led to contradictions, and that mathematicians would have to consider approaches to defining numbers other than those of naive set theory. This led to the development in the twentieth century of formal axiomatic sct theory.

Russell's paradox has encouraged others to look for paradoxes in our intuitive notions. Some of these paradoxes are both surprising and entertaining. Consider the Thomson Lamp paradox, discovered by the British philosopher James Thomson in 1970.[2] We pretend to have a perfect machine for turning a light on and off. First, we have the light on for one minute, after which it is turned off for one-half minute. Then it is on again for one-fourth minute and off for one-eighth minute. This continues with the light turned on or off after one-half the preceding time period. After two full minutes an infinite sequence of offs and ons will have occurred. After two minutes, will the light be on or off?

Another nice paradox is based on Russell's original paradox. Consider a town with only one barber. This barber shaves everyone who does not shave himself. Who shaves the barber? If the barber does not shave himself, then he does. If he does, then he should not.

Another clever paradox was discovered by the British librarian G. G. Berry. Suppose we classify all positive integers by the smallest number of syllables in English necessary to describe them. Hence, the number sev'en'teen requires three syllables while nine'ty-sev'en requires four syllables. Now we will consider the set of all integers that require at least nineteen syllables to describe them. Contained in this set is a smallest integer requiring nineteen syllables. However, we have just described this integer as the "least integer not describable using less than nineteen syllables," which is itself a description of eighteen syllables! Therefore, there is no least integer requiring nineteen syllables.[3]

Many of the paradoxes become paradoxes when a description is based on self-reference. Hence, we have the paradox involving sets containing themselves, descriptions describing themselves, and barbers shaving themselves. These paradoxes demonstrate on a conceptual level that we are still struggling with the primary notions that we use to formulate our ideas of numbers and our mathematics.

A second difficulty with the attempt to find axioms for all of mathematics involves the notion of completeness. It was assumed that a complete system of axioms, such as Peano's, should be strong enough to prove or disprove every statement formulated within the system. It might require years of work to decide for any particular statement, but, at least theoretically, it should be possible to prove each statement to be either true or false. This idea was destroyed by the work of Kurt Gödel (1906–1978) who proved in 1930 that within any system of axioms powerful enough to account for arithmetic, it is possible to formulate statements that cannot be proved. This is Gödel's famous *Incompleteness theorem*. It is this very theorem that demonstrated that it is not possible to prove if there is, or is not, a cardinal number between $\aleph_0$ and $\aleph_1$.

The discovery of the Incompleteness theorem dealt a blow to modern mathematics not unlike the blow dealt to the Pythagoreans by the discovery of the incommensurability of the diagonal. In a philosophical sense, it tells us that logical and mathematical analysis is insufficient to uncover all truths about numbers. Hence, there will always remain truths whose discovery will remain outside of mathematics proper.

BEYOND THE TWENTIETH CENTURY

In much of what we have covered we have been describing the work of pure mathematicians as it relates to our concept of number. However, the evolution of numbers is more complex than the theories developed by these professional mathematicians. Numbers are part of the very fabric of our culture. Therefore, while the theory of number has been evolving in the minds of mathematicians, the culture of numbers has been evolving throughout our society. If mathematics was only of interest to professional mathematicians and ignored by the rest of society, then the world would be a bleak place indeed. Our technology is driven by numbers and mathematics, and both play an intimate role in our daily existence.

One standard of how sophisticated our mathematics is becoming is the use of mathematics by average citizens. According to mathematics professor Jerry King, there are approximately fifty thousand professional mathematicians in this country if we count them by the memberships of the three national mathematical organizations. These fifty thousand mathematicians publish approximately twenty-five thousand mathematical research papers each year in fifteen hundred different journals.[4] King then points out that almost no one reads these twenty-five thousand papers. For the most part, they are read by the journal reviewers and the authors themselves. Therefore, while we have a great cluster of professional mathematicians, and these numerous mathematicians produce a great pile of work each year, this work does not have a direct impact on the rest of us.

But this is not to say that new discoveries in mathematics do not reach the general population, for, mysteriously, they do. What reaches us is the best of mathematics, that which is important to us in our nonmathematical lives. There have been several key phases in human history when the use of mathematics exploded among the general population. This is not to say that everyone in the society advanced his or her knowledge of numbers but that great segments of a population did so. One of the first times occurred when farming developed around 8500 B.C. in western Asia, when ordinary men and women had to learn to count and make accounts.

Another advance came when cities evolved around 3500 B.C., and rulers needed scribes and priests to make complex computations for everything from tracking the seasons to distributing the land and labor. This evolution was accelerated by the invention of writing around 3100 B.C. in Sumer.

Even though Greece experienced a Golden Age of mathematics and philosophy between 500 B.C. and A.D. 100, not much changed for ordinary citizens. The abacus and counting boards became popular as calculation aids, but the general population did not share in the riches of the mathematics being discovered.

This changed in 1455 when Johann Gutenberg demonstrated movable metal type for printing books. From that time on, it was possible to print large numbers of relatively cheap books for mass consumption. This helped promote the ideas of the Renaissance, including scientific discoveries and the Hindu–Arabic numerals imported from the Near East. This ushered in the technological age we enjoy today. Scientists and technicians could access the ideas of the mathematicians and apply their discoveries to everyday problems. In turn, such specialists began to appreciate mathematics for its aesthetic value.

The twentieth century has given us two great developments in mathematics that will again spread mathematics throughout the population. The first, of course, is the invention of computers.

MACHINE MATHEMATICS

Most professional mathematicians were slow to accept computers. While university engineers in the late fifties and early sixties were joyfully playing with the new gadgets, the mathematicians were shaking their heads and saying that machines could not do proofs, and proofs were what mathematicians did. Only the applied mathematicians realized their great utility, and the field of numerical analysis took a great leap forward. Over the last thirty years, most mathematicians have finally come around to using computers, even if many still feel that real mathematics is done with pencil and paper. Even the idea that computers could not do proofs has been challenged. The famous Four-Color theorem states that no more than four colors are needed to color any map in such a way that no two adjacent countries share the same color. The theorem was proposed in the nineteenth century but not proven until 1976 with the use of powerful computers. Proving the theorem with pencil and paper alone is not currently possible.

Yet, there are two other aspects of computers that are relevant to our discussion. First, computers are, in their very essence, binary number calculating machines. Everything in computers begins with, or is reduced to, the storage, addition, or subtraction of binary numbers. Therefore, at the very heart of computers are numbers.

However, the most important feature of computers for the general population is that the computer allows us to play with numbers. The importance of number play has already been mentioned and is often neglected in discussions about what mathematics is. Yet, the kind of activity that precedes the serious theorem-proving activity of the pure mathematician is the enthralling play with number ideas. Many mathematical discoveries were made because a mathematician was playing with numbers and suddenly noticed something different. The general population now has a great machine available with which to play with numbers. It is not that we just have a better pencil and paper, we have a machine that allows us to play with numbers in ways that were impossible without it. We can now ask ourselves number questions, and then use the machine to play while we search for the answers. Perhaps the answers will suggest themselves. This is an advantage that no other generation before us enjoyed. Only now, with the great distribution of home computers, does the average person have such power to investigate and study in a delightful way the basic nature of numbers. To carry out such investigations on a computer, the user must know a programming language. Therefore, young people should be encouraged to learn programming so that they might achieve full use of these wonderful machines.

Will the opportunity to carry out such activity by both mathematicians and nonmathematicians produce anything new? We have already experienced the first great discovery based on computer play—the theory of fractals.

LOOKING INTO GOD'S EYE

Most people have encountered fractals. They are those beautiful color pictures that we know are connected in some magical way to both mathematics and the weather. Specifically, fractals are complex geometric curves that do not change into simple forms under magnification and have a dimension between whole numbers. Two forms of this new developmental area exist: chaos theory in the sciences, and fractal theory in mathematics. The implications for both science and mathematics are phenomenal, and will change the face of our science and mathematics for centuries to come.

First we will consider the sciences. Mathematical models have been used for centuries to describe and predict physical events. The planets move in ellipses around the sun and cannon balls follow parabolas through the air. Science has been very successful in predicting such elementary occurrences. However, whenever the events become the least bit complex, our mathematical models seem to fizzle out. Why, with all their weather satellites and supercomputers, cannot the weather people predict if it is going to rain or be sunny tomorrow? We are not asking for the weather a month or even a week from now. We just want to know the next few days with any certainty. Yet, the meteorologists' ability to predict local weather conditions for a day or two is notoriously poor.

Why cannot we use our computers to predict the stock market and make us rich? What about the simple task of describing how water whirls around rocks in a stream? The truth is, many physical systems of nature simply cannot be described by classical mathematics. At one time it was believed that all we had to do was select the right set of equations and measure the proper initial conditions, and we would eventually solve all such problems. A number of the systems in nature that we study appear to be rather stable, and the course of events is not disturbed by minor bumps along the way. Take the orbit of the earth, for example. We do not expect the earth to move out of its orbit and leave the solar system because of an earthquake or an exploding volcano. Living organisms also appear quite resilient to change. Baby animals of the same species born to different environments still grow into the same general forms as adults. Surely, this is how all of nature works.

To tackle the difficult systems, such as weather, we evolved sets of equations that were so complex that a direct solution was not practical, and giant computers were used to get approximations. Still many of nature's systems have defied our best attempts at making predictions. Then scientists began to wonder if there were not systems in nature that were inherently unstable, that would veer off course at the least change in conditions. What they discovered were numerous systems extremely sensitive to initial conditions. In fact, they are so sensitive that it is not possible to measure the initial conditions with sufficient accuracy to make predictions. Enter chaos. We now realize that some aspects of the universe are so ultrasensitive that they appear to act in an utterly chaotic fashion. It does not matter how good our mathematics becomes, these systems will always manage to veer off in directions we cannot predict.

But all is not bad news. In discovering chaos, we have also discovered that a system that is sensitive and operates according to just a few simple laws can produce outrageously complex results. Hence, the very richness of nature is the result of the application of simple rules applied over and over in a sensitive system. The end result cannot be predicted from the simplicity of the rules and contains a richness of complexity that is unseen in those rules. This has implications for everything from how organisms grow from simple genetic codes to how the national economy operates. The discovery of chaos and sensitive systems was the direct result of trying to model such systems on computers.

Everything we have said so far appears related to applied mathematics. How could computers and systems in nature influence the clean, erudite world of pure mathematics? Astonishingly, our world of universals, as embodied in the number system, contains its own chaos. When we draw a computer picture of this chaos, we generate a fractal. How can this be? How can the precise mathematical principle of number contain pictures of chaos? To directly experience this chaos imbedded in numbers is to see right into the eye of God.

To generate a computer image of a mathematical fractal and see this chaos is so simple that anyone with a sound understanding of algebra and a home computer can produce one.[5] Let's take the simple polynomial, $x^3 + 8 = 0$, as an example. This equation contains an infinite amount of complex detail that was never imagined until just the last twenty years. Since the polynomial is of degree three, it has three solutions, x_1, x_2, and x_3. One solution is just the rational number -2. If we substitute -2 in for x_1 we get $(-2)^3 + 8 = -8 + 8 = 0$. The other two solutions are the complex numbers, $x_2 = 1 + \sqrt{3}i$ and $x_3 = 1 - \sqrt{3}i$. We can plot these three solutions on the complex number plane as is done in Figure 54.

A popular method of solving polynomials is the iteration approach. To use this method, we define a calculating algorithm so that when we guess at the solution and plug that guess into the algorithm, the result is a better guess. We then use this better guess in the algorithm to produce a second, and closer, guess. We continue in this manner until we are satisfied our answer is sufficiently close to the actual solution. It is interesting to note that any algorithm used for iteration has a kind of built-in self-reference principle that we encountered when discussing paradoxes. The solution to the algorithm is another solution to the algorithm, and so forth. Therefore, the iteration reflects back onto itself.

One kind of iteration method was developed by Isaac Newton (1642–1727), possibly the greatest combined scientist and mathematician who ever lived. In this method we have the following algorithm for iterating our solutions for $x^3 + 8 = 0$, where each guess is shown as an E.

$$E_2 = E_1 - \frac{E_1{}^3 + 8}{3E_1{}^2}$$

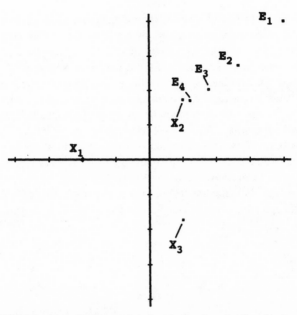

FIGURE 54. The three solutions to $x^3 + 8 = 0$ plotted on the Gaussian (or complex) plane as $x_1, x_2,$ and x_3. Also plotted are the first four values ($E_1, E_2, E_3,$ and E_4) we get when using Newton's method to find x_2.

Let's begin with our first estimate equal to the complex number $4 + 4i$. This E_1 has been plotted in Figure 54. When plotting our complex numbers we will designate them as ordered pairs, such as (4,4) for $4 + 4i$. Now, since E_1 is closest to our solution x_2, we might assume that beginning with this estimate we will end up generating successive guesses that move ever closer to x_2. This is, in fact, just what happens. Substituting $4 + 4i$ into the above algorithm produces an $E_2 = 2.67 + 2.75i$, which has also been plotted on Figure 54. Now, substituting E_2 into the algorithm produces E_3, which in turn produces E_4. Now we are quite close to the correct solution of $1 + \sqrt{3}i$ which is approximately $1 + 1.73i$. Everything so far is working as expected. Using Newton's method we begin with a guess close to x_2 and end with a sequence of estimates that converge on x_2.

We might assume that whenever our first estimate, E_1, is closest to a particular solution, x_i, then the following sequence of estimates would converge on that x_i. Figure 55 shows the complex number plane divided by heavy lines into three areas around the three solutions. We would assume all first estimates around x_2 should converge to x_2, and similarly with the other two areas—first estimates always converging to the closest solution. The only question might be regarding the points falling exactly between two solutions. This is the question that John Hubbard's first-year calculus students asked him back in 1976.[6] Do the initial guesses always go to the closest solution? He was not sure of the answer and told them he would let them know the following week.

Hubbard could not find a good answer in a week. He had to go to his computer and begin to play. His play led to his discovery that something

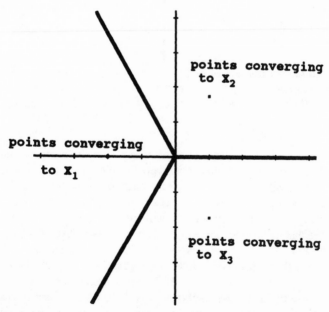

FIGURE 55. The complex plane divided into three equal areas. When using Newton's method of solution, the logical assumption is that those points closest to each solution point converge to that point. Do they?

very strange was occurring in the complex plane. The original assumption about guesses converging to the closest solution had made sense and was logical, but it was also dead wrong. Look at what happens in Figure 56 when we show the source of all first estimates that lead to x_1! We do not get a clear demarcation between the different solutions, we get a very complex picture—we get a fractal. What happens if we take a small area of complex detail and magnify it? The detail does not resolve itself, but continues (Figure 57). That is, no matter how much we magnify the image between the various solutions that have detail, we never get a clear and simple demarcation. The rich detail of the fractal continues on, forever!

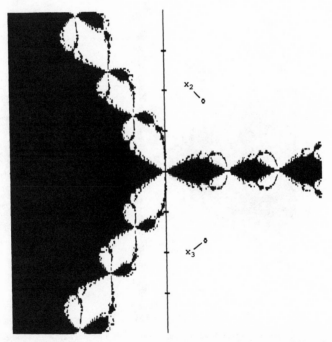

FIGURE 56. Testing points on the complex plane with Newton's method yields a fractal image. The black areas in the figure are those first estimates in Newton's method that lead to x_1 (not shown). The white areas are estimates that lead to either x_2, or x_3. Considering the black areas to the right of the vertical axis, we may ask: Why do first estimates that are much closer to either x_2 or x_3 lead to x_1?

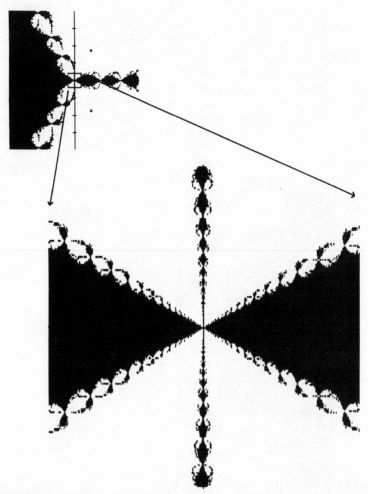

FIGURE 57. Magnifying the fractal generated by $x^3 + 8 = 0$ always yields more detail and does not resolve it into simpler forms. The small box in the original fractal has been magnified a hundred times to show this replication of complexity.

Others, besides Hubbard, had been investigating the strange pictures in the complex plane. The first to do so were the two French
mathematicians, Gaston Julia and Pierre Fatou, who had to do their
calculating without the aid of a computer during World War I. But their
work, and their resulting beautiful images, called Julia sets, were largely
ignored. During the 1970s Benoit Mandelbrot began studying Julia sets
and in 1979 discovered the granddaddy of complex plane fractals, the
Mandelbrot set.[7]

The remarkable thing about fractal images, such as the one in Figure
56, is that they do not rely in any way upon the chaos of a real-world
system. The only concepts used come directly from pure mathematics. In
Figure 56 we are looking at a fundamental characteristic of our number
system. Why does this detail exist? Every polynomial of degree two or
greater will generate a fractal image using the same general technique.
We used $x^3 + 8$ only because it is so simple. If we invented all the ideas
relating to mathematics, how did we invent the image in Figure 56?

An experience, not unlike listening to a great symphony, is to sit
before a computer monitor and watch a color image of a polynomial
fractal come alive. As the detail evolves, an electrical charge rushes
thróugh your body as you realize that you are looking through some great
rip in the fabric of being to stare at the essence of existence. Fractals are
another example of a discovery that leads mathematicians to believe in the
absolute existence of mathematical objects. In his excellent book, *Chaos:
Making a New Science*, James Gleick ventures to say:

> The Mandelbrot set . . . exists. It existed before . . . Hubbard and
> Dauady understood its mathematical essence, even before Mandelbrot
> discovered it. It existed as soon as science created a context—a
> framework of complex numbers and a notion of iterated functions. Then
> it waited to be unveiled. Or perhaps it existed even earlier, as soon
> as nature began organizing itself by means of simple physical laws,
> repeated with infinite patience and everywhere the same.[8]

FERMAT'S LAST THEOREM

Before we leave the accomplishments of the twentieth century, we
cannot fail to mention a recent event that may rank as one of the most

outstanding achievements in mathematics. You will remember that Pythagorean numbers are whole number triplets that satisfy the following equation: $a^2 + b^2 = c^2$. We know that there are an infinite number of such triplets, and they were known by the ancient Babylonians. It is a natural extension of Pythagorean numbers to ask if any three natural numbers exist such that their cubes have the following relationship: $a^3 + b^3 = c^3$.

Pierre de Fermat was born in 1601 in Beaumont-de-Lomagne, France. He is considered by some to be the seventeenth century's best mathematician, and he was certainly one of the world's greatest amateur mathematicians, doing important work on the foundation of calculus, independently from Descartes inventing analytic geometry, and establishing the foundations of modern number theory.[9]

Fermat, in 1637, while studying a copy of the second book of Diophantus's *Arithmetic*, jotted on the margin of said text:

> On the contrary, it is impossible to separate a cube into two cubes, a fourth power into two fourth powers, or, generally, any power above the second into two powers of the same degree: I have discovered a truly marvelous demonstration which this margin is too narrow to contain.[10]

Here, Fermat has answered our question and much more. Not only is it impossible to find three natural numbers that satisfy $a^3 + b^3 = c^3$, but it is impossible to find three numbers to satisfy $a^n + b^n = c^n$ whenever n is larger than two. This simple assertion is known as Fermat's Last Theorem. However, Fermat never wrote down his proof, and many think he was mistaken to believe he knew one, for the tools necessary to prove his theorem were not available in 1637. Yet, who can be sure? What is known is that from 1637 until 1993, a period of 356 years, no one could prove Fermat's Last Theorem, and countless professional and amateur mathematicians tried. I can remember as a graduate student at the University of Utah that my mathematics professors warned us not to fool around trying to prove Fermat's Last Theorem, for too many students had become obsessed with it, causing their studies to suffer. Not all of us listened to this warning.

Over three days, from June 21 to June 23, 1993, a shy English mathematician from Princeton University, Andrew Wiles, presented three one-hour lectures on "Modular Forms, Elliptic Curves, and Galois

Representations."[11] From the first day, those in the audience guessed where Wiles was headed. At the end of the third lecture, Wiles wrote Fermat's Last Theorem on the board and the audience broke out in wild applause. Finally, after 356 years, someone had presented a proof. But, will his proof stand up under close scrutiny? We will have to wait to find out, for Wiles must still publish his proof in a journal so that other mathematicians can check his work. However, those attending his lecture feel positive about the chances, and believe he has succeeded. If so, Andrew Wiles is guaranteed his place in history books for cracking one of the most difficult problems in mathematics, and his proof will rank as one of the single greatest mathematical achievements of the twentieth century.

HOW FAR CAN WE GO?

Now that we have our computers and our mathematics books, how far can we go in discovering new mathematics? If our brains are such that the average intelligence is roughly one hundred points on the I.Q. scale, how far can our native intelligence take us in our pursuit of ever more elegant mathematics? Of course, if we were visited by those aliens from the center of the galaxy, and they turned out to have I.Q.'s of two thousand, we might feel somewhat intimidated by their thinking prowess. Could we dare to believe we might understand their mathematics? Would we be like chickens or dogs listening to lectures on the theory of calculus but understanding nothing except that someone was speaking to us? Is there a limit to where we can go, simply because of the limiting architecture of our brains?

Suppose that one day we unlock all the keys to our genetic code. Suppose further that we can manipulate that code to give our children desirable programmed characteristics. Could we then deliberately program children with I.Q.'s of two hundred, or four hundred, or even two thousand? What kind of mathematics would they discover? Whatever it would be, those of us in the old generation would probably never understand it. But, we would know that it must be grand and elegant.

End Notes

INTRODUCTION

1. Kathleen Freeman, *Ancilla to the Pre-Socratic Philosophers* (Cambridge, MA: Harvard University Press, 1966), p. 74.
2. Plato, *The Dialogues of Plato*, trans. B. Jowett (New York: Random House, 1937) *The Republic*, VII, 525.

CHAPTER 1

1. Karl Menninger, *Number Words and Number Symbols* (New York: Dover Publications, 1969), p. 33.
2. Erich Harth, *Windows on the Mind: Reflections on the Physical Basis of Consciousness* (New York: William Morrow and Company, 1982), p. 101.
3. Paul Glees, *The Human Brain* (Cambridge: Cambridge University Press, 1988), p. 37.
4. *Mathematical Disabilities* (Gérard Deloche and Xavier Seron, eds.) (Hillsdale, New Jersey: Lawrence Erlbaum Associates, 1987).
5. Paul D. MacLean, *The Triune Brain in Evolution* (New York: Plenum Press, 1990), p. 549; Deloche and Seron, p. 140.

CHAPTER 2

1. Roger Lewin, *Bones of Contention: Controversies in the Search for Human Origins* (New York: Simon and Schuster, 1987), p. 108.

281

2. David Lambert, *The Field Guide to Early Man* (New York: Facts on File, 1987), pp. 98–105.
3. Ibid., p. 106.
4. David Eugene Smith, *History of Mathematics* (New York: Dover Publications, 1951), p. 6.
5. Paul D. MacLean, *The Triune Brain in Evolution* (New York: Plenum Press, 1990), p. 555.
6. Richard E. Leakey, *Origins* (New York: E.P. Dutton, 1977), p. 205.
7. Karl Menninger, *Number Words and Number Symbols*, p. 35.
8. Graham Flegg, *Numbers Through the Ages* (London: MacMillan Educations LTD, 1989), p. 7.
9. Graham Flegg, *Numbers: Their History and Meaning* (New York: Schocken Books, 1983), p. 19.
10. Flegg, *Numbers Through the Ages*, p. 9.
11. Flegg, *Numbers: Their History and Meaning*, p. 24.
12. Menninger, p. 11.
13. Flegg, *Numbers: Their History and Meaning*, p. 11.
14. Menninger, p. 32.
15. Leakey, p. 162.
16. Flegg, *Numbers Through the Ages*, p. 37.
17. Ibid., p. 11.

CHAPTER 3

1. Graham Flegg, *Numbers: Their History and Meaning*, p. 7.
2. H. Kalmus, "Animals as Mathematicians," *Nature 202* (June 20, 1964), p. 1156.
3. Levi Leonard Conant, "Counting," in *The World of Mathematics, Vol. I* (James R. Newman, ed.) (New York: Simon and Schuster, 1956), p. 433.
4. Donald R. Griffin, *Animal Thinking* (Cambridge, MA: Harvard University Press, 1984), p. 204.
5. O. Koehler, "The Ability of Birds to 'Count'," in *The World of Mathematics, Vol. I* (James R. Newman, ed.) (New York: Simon and Schuster, 1956), p. 491.
6. Conant, p. 434.
7. Guy Woodruff and David Premack, "Primate Mathematical Concepts in the Chimpanzee: Proportionality and Numerosity," *Nature 293* (October 15, 1981), p. 568–570.
8. Phone conversation with Kenneth S. Norris, retired professor of natural history at the University of California–Santa Cruz, Nov. 19, 1992.
9. Menninger, *Number Words and Number Symbols*, p. 11.
10. John McLeish, *Number* (New York: Fawcett Columbine, 1991), p. 7.
11. David Caldwell and Melba Caldwell, *The World of the Bottle-Nosed Dolphin* (New York: J. B. Lippincott Co., 1972), p. 17.

12. Carl Sagan, *Mind in the Waters* (Joan McIntyre, ed.) (New York: Charles Scribner's Sons, 1974), p. 88.

CHAPTER 4

1. David Eugene Smith, p. 37.
2. Denise Schmandt-Besserat, *Before Writing, Vol. I: From Counting to Cuneiform* (Austin, TX: University of Texas Press, 1992), p. 7.
3. Ibid., p. 6.
4. Ibid., p. 190.
5. Mortimer Chambers, Raymond Grew, David Herlihy, Theodore Rabb, and Isser Woloch, *The Western Experience: To 1715* (New York: Alfred A. Knopf, 1987), p. 7.
6. Schmandt-Besserat, p. 114.
7. Ibid., p. 199.
8. Carl B. Boyer, *A History of Mathematics* (New York: John Wiley and Sons, 1968), p. 33.
9. H. L. Resnikoff and R. O. Wells, Jr., *Mathematics in Civilization* (New York: Dover Publications, 1973), p. 76.
10. David Eugene Smith, p. 43.
11. Boyer, p. 22.
12. David Eugene Smith, p. 43.
13. Boyer, p. 12.
14. Morris Kline, *Mathematical Thought from Ancient to Modern Times, Vol. 1* (New York: Oxford University Press, 1972), p. 16.
15. Lucas Bunt, Phillip Jones, and Jack Bedient, *The Historical Roots of Elementary Mathematics* (New York: Dover Publications, 1976), p. 37.

CHAPTER 5

1. McLeish, p. 53.
2. David Eugene Smith, p. 23.
3. Menninger, p. 452.
4. Boyer, p. 220.
5. McLeish, p. 70.
6. Ibid., p. 24.
7. Stuart J. Fiedel, *Prehistory of the Americas* (Cambridge: Cambridge University Press, 1987), p. 282.
8. Ibid., p. 281.
9. The *Codex Dresdensis* in Dresden, the *Codex Tro-Cortesianus* in Madrid, and the *Codex Peresianus* in Paris.

10. Bunt *et al.*, p. 226.
11. Thomas Crump, *The Anthropology of Numbers* (New York: Cambridge University Press, 1990), p. 46.
12. Jacques Soustelle, *Mexico* (New York: World Publishing Company, 1967), p. 125.
13. Fiedel, p. 335.

CHAPTER 6

1. Chambers *et al.*, p. 40.
2. Menninger, p. 272.
3. Ibid., p. 299.
4. Kline, p. 28.
5. David Eugene Smith, p. 64.
6. The two different positions are illustrated by Smith, p. 71; and Boyer, p. 52.
7. Michael Moffatt, *The Ages of Mathematics: Vol. I, The Origins* (New York: Doubleday and Company, 1977), p. 96.
8. Boyer, p. 60.
9. Bunt *et al.*, p. 83.
10. Aristotle, *The Basic Works of Aristotle*, trans. J. Annas (Richard McKeon, ed.) (New York: Random House, 1941); *The Metaphysics*, 986a, lines 1–3 and 15–18, Oxford University Press.
11. Ibid., 1090a, lines 20–25.
12. Two different visual proofs come from Stuart Hollingdale, *Makers of Mathematics* (London: Penguin Books, 1989), p. 39; and Eric Temple Bell, *Mathematics: Queen and Servant of Science* (New York: McGraw-Hill, 1951), p. 190.
13. Kline, p. 33.
14. Moffatt, p. 92.
15. Bunt *et al.*, p. 86.

CHAPTER 7

1. McLeish, p. 115.
2. Kline, p. 184.
3. David Eugene Smith, p. 157.
4. Menninger, p. 399.
5. McLeish, p. 122.
6. Kline, p. 184.
7. Bunt *et al.*, p. 226.
8. Ibid., p. 227.
9. Menninger, p. 425.

10. Ibid., p. 432.
11. Ibid., p. 400.
12. Hollingdale, p. 109.
13. Jane Muir, *Of Men and Numbers* (New York: Dodd, Mead, and Company, 1961), p. 235.

CHAPTER 8

1. Freeman, p. 14.
2. Ibid., p. 19.
3. Aristotle, *The Basic Works of Aristotle, Physics, Book III*, 204b, lines 2–9.
4. Freeman, p. 75.
5. Aristotle, *Physics, Book III*, 206b, lines 31–32.
6. Ibid., 204b, lines 6–8.
7. Ibid., 206a, line 26; 206b, line 13.
8. Ibid., 239b, lines 14–18.
9. Plato, *The Dialogues of Plato*, trans. B. Jowett (New York: Random House, 1937), Timeaus, lines 25, 52.
10. Rudy Rucker, *Infinity and the Mind* (New York: Bantam Books, 1982), p. 3.
11. Thomas Hobbes, *Leviathan: Parts I and II* (New York: Bobbs-Merrill Company, 1958), p. 36.
12. Thomas Hobbes, "Selections from the De Corpore," in *Philosophers Speak for Themselves: From Descartes to Locke* (T. V. Smith and Marjorie Grene, eds.) (Chicago: University of Chicago Press, 1957), p. 144.
13. René Descartes, "Meditations on First Philosophy," in *Philosophers Speak for Themselves: From Descartes to Locke*, p. 78.
14. Hollingdale, p. 359.
15. Rucker, p. 88.
16. Euclid, *Elements, Book III* (New York: Dover Publications, 1956), Sec. 14.

CHAPTER 9

1. A well-ordered set is a simply ordered set such that every subset contains a first element. See Zermelo's axiom of choice.
2. Sir Thomas Heath, *A History of Greek Mathematics, Vol I* (Oxford, England: The Clarendon Press, 1960), p. 385.
3. The English translation of this work can be found in Richard Dedekind, *Essays on the Theory of Numbers* (La Salle, IL: Open Court Publishing Company, 1948).
4. Ibid., p. 6.
5. Ibid., p. 12.

6. Ibid., p. 13.
7. Ibid., p. 15.
8. Boyer, p. 307.
9. Ibid., p. 348.
10. Richard Preston, "Profiles: The Mountains of Pi," *The New Yorker* (March 2, 1992), p. 36.

CHAPTER 10

1. Hollingdale, p. 275.
2. Boyer, p. 361.
3. Muir, p. 217.
4. Quoted in Sherman K. Stein, *Mathematics: The Man-Made University*, (New York: W. H. Freeman and Company, 1963), p. 252.
5. Ibid., p. 253.
6. Joseph Warren Dauben, *Georg Cantor: His Mathematics and Philosophy of the Infinite* (Princeton, NJ: Princeton University Press, 1979), p. 50.
7. Leo Zippin, *Uses of Infinity* (Washington, D.C.: The Mathematical Association of America, 1962), p. 56.

CHAPTER 11

1. Kline, p. 143.
2. Ibid., p. 253.
3. Eric Temple Bell, *Men of Mathematics* (New York: Simon and Schuster, 1965), p. 35.
4. Hollingdale, p. 126.
5. Bell, *Men of Mathematics*, p. 43.
6. Muir, p. 172.
7. Dauben, p. 54.
8. Ibid., p. 55.
9. Hollingdale, p. 337.

CHAPTER 12

1. E. Kamke, *Theory of Sets* (New York: Dover Publications, 1950), p. 47.
2. Rucker, pp. 48–50.
3. Dauben, p. 232.

4. Rucker, p. 276.
5. Ibid., pp. 281–285.
6. Muir, p. 237.
7. Dauben, p. 285.
8. Ibid., p. 243.

CHAPTER 13

1. Steven B. Smith, *The Great Mental Calculators* (New York: Columbia University Press, 1983).
2. Darold A. Treffert, *Extraordinary People* (New York: Harper & Row Publishers, 1989).
3. Steven B. Smith, p. 97.
4. Ibid., p. 289.
5. Ibid., p. 245.
6. Oliver Sacks, *The Man Who Mistook His Wife for a Hat* (New York: Harper Perennial, 1985), p. 203.
7. Treffert, p. 41.
8. W. W. Rouse Ball, "Calculating Prodigies," in *The World of Mathematics, Vol. 1*, p. 467.
9. Steven B. Smith, p. xv.
10. Treffert, p. 220.
11. Ibid., p. 222.
12. A. E. Ingham, *The Distribution of Prime Numbers* (Cambridge: Cambridge University Press, 1990), p. 3.

CHAPTER 14

1. Bertrand Russell, *The Problems of Philosophy* (London: Oxford University Press, 1959), p. 12.
2. George Berkeley, "Three Dialogues Between Hylas and Philonous," in *Philosophers Speak for Themselves: Berkeley, Hume, and Kant*, pp. 1–95.
3. Russell, p. 98.
4. Jerry P. King, *The Art of Mathematics* (New York: Plenum Press, 1992), p. 29.
5. Ibid., p. 43.
6. Ibid., p. 139.
7. W. H. Werkmeister, *A Philosophy of Science* (Lincoln, NE: University of Nebraska Press, 1940), p. 141.

CHAPTER 15

1. Boyer, p. 645.
2. E. J. Borowski and J. M. Borwein, *The HarperCollins Dictionary of Mathematics* (New York: HarperCollins Publishers, 1991), p. 589.
3. Ibid., p. 49.
4. King, p. 6.
5. Roger T. Stevens, *Fractal: Programming in Turbo Pascal* (Redwood City, CA: M&T Publishing, 1990).
6. James Gleick, *Chaos: Making a New Science* (New York: Viking Penguin, 1987), p. 217.
7. Ibid., p. 222.
8. Ibid., p. 239.
9. Bell, *Men of Mathematics*, p. 57.
10. Ibid., p. 71.
11. Michael D. Lemonick, "*Fini* to Fermat's Last Theorem," *Time* (5 July 1993), p. 47.

Glossary

abacus: an ancient calculating device consisting of beads strung on rods mounted in a wooden frame. Calculations are performed by moving the beads along the rods.

Absolute Infinite: entity that is identified with a collection of all infinities, sometimes associated with God and designated as Ω (capital omega). A concept that cannot be understood rationally but only mystically.

absolute value: a positive value of a number regardless of the number's original sign.

Alexandrian number system: the Greek numbering system based on twenty-seven different numerals and used predominately after 100 B.C. Also called the Ionic system.

algebraic number: a number that is a solution to a polynomial whose coefficients are all rational numbers.

analytic geometry: the geometry where positions are represented by number coordinates and algebraic methods apply.

Attic number system: an early, written Greek number system based on six primary numerals and used until approximately 100 B.C. Also called the Herodianic system.

body-counting: an extension of finger-counting where different body parts represent natural numbers.

Brahmi numerals: Hindu numerals for the numbers 1 through 9, which were used as early as the third century B.C. These symbols were to eventually lead to the modern Hindu–Arabic numerals used today.

cardinal number: a number specifying how many elements are in a set.

Cartesian coordinate system: the use of two perpendicular number lines to identify every point in the plane with two real numbers.

closed numbers: a set of numbers is closed for an operation if every application of that operation on numbers from the set yields another number in the set.

cluster point: another name for a limit, especially when more than one limit is involved.

complex number: a number of the form (a,b) or $a + bi$ where both a and b are real numbers and i is $\sqrt{-1}$. Real numbers and imaginary numbers are both subsets of the complex numbers.

composite number: a natural number that can be evenly divided by more numbers that 1 and itself.

concrete counting: when the objects being counted are mapped one-to-one with counting tokens or symbols.

continued fraction: a number consisting of an integer plus a fraction such that the denominator of the fraction is also an integer plus a fraction of the same kind. Every irrational number can be represented as an infinite continued fraction.

constructivism: the belief that mathematical objects do not exist independently of human minds.

continuum hypothesis: the hypothesis that there exists a transfinite number between $\aleph_0$ and $\aleph_1$. The question is undecidable.

convergence: state of an infinite sequence of numbers or number series that approaches a limit.

coordinates: the two real numbers associated with each point in the plane in analytic geometry.

countable set: an infinite set that can be put into a one-to-one mapping with the natural numbers. Also called "enumerable."

counting: finding the cardinal number of a set.

counting board: an ancient counting device consisting of a marked board and counting sticks or pebbles.

dense: the property that between every two numbers there exists another number. The rational, irrational, real, and complex numbers are all dense.

denumerable set: a set that can be mapped one-to-one with the natural numbers. See countable set.

digit: any one of the ten numerals 0, 1, 2, 3, 4, 5, 6, 7, 8, and 9 of the Hindu–Arabic number system.

divergence: state of an infinite sequence of numbers or number series that has no bound or limit.

element: one particular item or member of a set.

Euclidean geometry: the geometry developed by Euclid and satisfying the parallel postulate that says: Given a line and a point not on the line, one and only one other line can pass through the point that is parallel to the original line.

exponent: a number written as a superscript to a second (base) number and indicating the power to which the second number is to be raised.

extendable cardinal number: the largest cardinal number known at present.

factorial: the product of all the natural numbers less than and equal to a specific natural number and indicated with the symbol "!" placed after the specific natural number. For example, $5! = 1 \cdot 2 \cdot 3 \cdot 4 \cdot 5 = 120$.

false position: a method of solving equations using a guess substituted into the equation to generate a proportional adjustment for an improved guess.

Fibonacci sequence: the number sequence beginning with 1 where each succeeding term is the sum of the previous two terms. The first seven terms of the sequence are 1, 1, 2, 3, 5, 8, 13, . . .

finger-counting: using fingers to map onto a set of objects to be counted.

five-counting (5-counting): an early number system with a base of five.

fractal: a complex geometric curve that does not change into simple forms under magnification and generally has a dimension that is between whole numbers.

Gobar numerals: the numerals used by the Arabs during the ninth century A.D. and transcribed from the earlier Indian Brahmi numerals.

golden mean or golden ratio: the ratio of $(1 + \sqrt{5})$ to 2 or the fraction $(1 + \sqrt{5})/2$. Discovered by the Greeks and found in numerous mathematical relationships.

hieratic writing: early Egyptian writing used by scribes to conduct daily record-keeping.

hieroglyphics: early Egyptian writing used for formal occasions and on monuments.

Hindu–Arabic number system: the dominant number system used today based on the ten unique symbols 0, 1, 2, 3, 4, 5, 6, 7, 8, and 9, and using a position-value system.

hypercomplex number: a number generated by extending the concept of number to dimensions beyond the two-dimensional complex numbers. See quaternions.

hyperinaccessible transfinite number: a transfinite number that is so inaccessible that its inaccessibility cannot be defined beginning with smaller transfinite numbers. See inaccessible transfinite number.

hypotenuse: in a triangle, the side opposite the right angle.

imaginary number: a number on the vertical axis of the complex number plane; a number in the form ai where a is a real number and i is $\sqrt{-1}$.

inaccessible transfinite number: a transfinite number that cannot be defined in terms of smaller transfinite numbers.

incommensurable: when two magnitudes cannot be expressed as the ratio of two whole numbers.

indeterminate equations: equations with an infinite number of solutions, for example, the equation $x + y = 7$. Such equations are useful in the study of numerous physical systems.

indirect method of proof: a method of proof discovered by various ancient societies where one assumes the opposite of what is to be proven and then shows a contradiction. Also called *reductio ad absurdum*.

infinite: without bounds, unbounded, not finite. For a set, able to be mapped onto a proper subset of itself, for example, the natural number map onto the squares of natural numbers.

integer: one of the set of numbers consisting of the positive and negative natural numbers plus zero.

intuitionism: the belief that mathematics should not deal with infinite sets and that only those proofs involving finite steps or constructions should be admitted.

irrational number: a number on the real number line that cannot be expressed as the ratio of two whole numbers.

limit: the sequence of terms $A_1, A_2, A_3, \ldots A_n, \ldots$ has a limit L if for any positive value ϵ there exists a number N such that for all $n > N$, the absolute value of $L - A_n < \epsilon$.

linear equation: an equation where all the unknowns have exponents equal to 1.

Liouville numbers: transcendental numbers of the following form:

$$a_1/10^{1!} + a_2/10^{2!} + a_3/10^{3!} + \ldots$$

where the a's are integers in the range of 0 to 9.

logarithm: the exponent to which a number called a base must be raised to obtain a given number. The logarithm of n with the base of a is written as $\log_a n$. For example, $\log_{10} 100 = 2$, or $10^2 = 100$.

logistic: the ancient Greek art of calculation, considered a trade rather than an intellectual pursuit.

method of exhaustion: a technique developed by the Greeks to find the area within curved geometric shapes. The technique finds successively better approximations by computing the areas of rectangles and triangles within the curve.

neo-2-counting: an early form of counting using number words equivalent to "one" and "two" plus conjunctions of these two symbols and involving addition, subtraction, and multiplication.

non-Euclidean geometry: any geometry based on substituting a different postulate for Euclid's parallel postulate. See Euclidean geometry.

nonpositional number system: any number system where the position of a specific numeral does not help determine the numeral's value. Hence, the numerals may be written in any order without changing the number's value.

normal number: a number whose decimal expansion contains an equal proportion of all ten digits from 0 through 9. An absolutely normal number is one whose decimal expansion under all bases contains an equal proportion of each digit.

number field: any set of real or complex numbers such that the sum, difference, product, and quotient (except for zero) is another number in the set. Therefore a number field is closed under the four operations of arithmetic.

number line: an infinite line where each point is associated with one and only one real number.

number-number: the decimal number formed by writing each natural number in succession, for example, 0.12345678910111213 . . .

number sequence: a set of numbers: $A_1, A_2, A_3, \ldots A_n, \ldots$ forms a sequence of numbers if the numbers are well ordered, that is, if the subscripts are in the order of the natural numbers.

number series: a collective sum of a set of numbers.

number theory: the formal study of the natural numbers and their relationships.

numeral: the written symbol representing a number or representing a digit of a number.

numerical analysis: the use of computers and computing algorithms to approximate the solutions to complex problems.

one-to-one mapping: assignment of exactly one and only one element from a set (e.g., number words) to each element in a second set (e.g., fingers).

ontology: the study of the meaning of existence.

ordinal number: a number that specifies the relative position of an element in a set.

origin: the point on the number line associated with the number zero, or the point in the complex plane where the two axes intersect.

paradox: an argument that derives self-contradictory conclusions by valid deduction from intuitively acceptable premises.

perfect number: a natural number that is the sum of its divisors. Six is the first perfect number since $6 = 1 + 2 + 3$, and 28 is the second since $28 = 1 + 2 + 4 + 7 + 14$.

polynomial equation: an equation of one or more unknowns raised to powers and multiplied by numbers called coefficients. A polynomial with one unknown, x, has the general form $a_0x^n + a_1x^{n-1} \ldots + a_{n-1}x + a_n = 0$.

positional number system: a number system that combines the value of a numeral with its position within the number to give its final value. The modern Hindu–Arabic number system is a positional system.

prime number: a natural number that can only be evenly divided by itself and the number 1.

projective geometry: that branch of mathematics concerned with those properties of geometrical figures that do not change when the figures are projected onto a different space.

proof: a sequence of logical steps that establishes the truth of a conclusion based on a set of axioms. The first fully articulated proofs were developed by the Greeks to prove certain geometric relationships.

Pythagorean numbers: any set of three natural numbers that satisfy the Pythagorean theorem. For example, the triplet 3, 4, and 5 are Pythagorean numbers since $3^2 + 4^2 = 5^2$.

Pythagorean theorem: the square of the length of the hypotenuse of a right triangle is equal to the sum of the squares of the lengths of the adjacent sides, or $a^2 + b^2 = c^2$ where c is the length of the hypotenuse and a and b are the lengths of the adjacent sides.

quaternions: complex numbers of the form $a + bi + cj + dk$ where a, b, c, and d are real numbers and i, j, and k are hypercomplex numbers such that $i^2 = j^2 = k^2 = ijk = -1$.

quipu: a collection of knotted strings used by the Incas of the New World to record possessions and transactions.

rational number: any number that can be expressed as the ratio of two nonzero integers.

real number: the numbers associated with all the points on the number line; the union of the algebraic and transcendental numbers.

rhetorical algebra: an ancient form of algebra where problems were stated in ordinary language without precise symbolism.

root: a number that satisfies an equation; that is, when substituted into the equation for an unknown, both sides of the equal sign are equal in value.

Russian peasant method: a method originally used by the early Egyptians to multiply numbers by successively doubling one number and then adding the appropriate multiples; adopted later by Central Europeans.

set: a collection of items or elements.

sexagesimal number system: a number system based on sixty, as opposed to our Hindu–Arabic system, which is based on ten.

simply ordered: a set of numbers is simply ordered when the following two conditions hold for any three numbers of the set, x, y, and z: (1) $x = y$ or $x < y$ or $y < x$ and (2) if $x < y$ and $y < z$ then $x < z$. The real numbers are simply ordered while the complex numbers are not.

simultaneous linear equations: two or more equations where each equation contains one or more unknowns and every unknown is raised to an exponent of 1.

square number: a natural number that is the square of another natural number. For example, 4 and 9 are square numbers because they are the squares of 2 and 3.

stick-counting: a form of counting that does not require the use of language. Counting objects, such as sticks or pebbles, are mapped one-to-one onto the set of objects being counted.

subitizing: the immediate awareness of the manyness of a set of objects.

symbolic algebra: algebra employing well-defined symbols rather than ordinary language.

syncopated algebra: algebra that is midway between rhetorical algebra and symbolic algebra; the use of a mix of symbols and ordinary words to make algebraic statements.

tag number: a number used in place of a name.

tally stick: a stick, usually split lengthwise, marked with notches to record financial transactions. When the two parts are placed together, the notches match.

theorem: a statement or formula that is deduced from a set of axioms and/or other theorems.

token: in counting, a small clay figurine used in western Asia between 8500 and 3000 B.C. for recording the cardinal number of sets of objects.

transcendental number: a real number that is not the root or solution to any polynomial with rational coefficients. A real number that is not algebraic.

transfinite number: the cardinal or ordinal number of an infinite set.

trigonometry: the study of the relationships between the lengths of the sides of a triangle and the measures of its interior angles.

two-counting (2-counting): an early form of counting using number words equivalent to "one" and "two" plus additive conjunctions of these two symbols.

uncountable: an infinite set whose elements cannot be mapped one-to-one with the natural numbers.

undecidable: within a formal mathematical or logical system, a statement that cannot be proved or disproved based on the axioms used in the system.

unit fraction: a fraction of the form $1/n$; a fraction whose numerator is 1.

vigesimal number system: a number system based on twenty, as opposed to the Hindu–Arabic system, which is based on ten.

well-ordered: a set is well-ordered if for every subset, including the set itself, there is a first element. The empty set, $\emptyset$, is considered well-ordered.

zero: the cardinal number of the empty set; the symbol that has no value but is used for a placeholder in a positional number system; the numeral 0.

Bibliography

Aristotle, *The Basic Works of Aristotle* (Richard McKeon, ed.). New York: Random House, 1941.

Bell, Eric Temple, *The Magic of Numbers*. New York: Dover Publications, 1974.

Bell, Eric Temple, *Men of Mathematics*. New York: Simon & Schuster, 1965.

Bell, Eric Temple, *Mathematics: Queen and Servant of Science*. New York: McGraw-Hill Book Company, 1951.

Borowski, E. J., and Borwein, J. M., *The HarperCollins Dictionary of Mathematics*. New York: HarperCollins Publishers, 1991.

Boyer, Carl B., *A History of Mathematics*. New York: John Wiley and Sons, 1968.

British Museum (Natural History), *Man's Place in Evolution*. London: Cambridge University Press (undated).

Bunt, Lucas; Jones, Phillip; and Bedient, Jack, *The Historical Roots of Elementary Mathematics*. New York: Dover Publications, 1976.

Burgess, Robert F., *Secret Languages of the Sea*. New York: Dodd, Mead, and Company, 1981.

Caldwell, David, and Caldwell, Melba, *The World of the Bottle-Nosed Dolphin*. New York: J. B. Lippincott Company, 1972.

Calvin, William H., *The Ascent of Mind: Ice Age Climates and the Evolution of Intelligence*. New York: Bantam Books, 1990.

Chambers, Mortimer; Grew, Raymond; Herlihy, David; Rabb, Theodore; and Woloch, Isser, *The Western Experience: to 1715*. New York: Alfred A. Knopf, 1987.

Crump, Thomas, *The Anthropology of Numbers*. New York: Cambridge University Press, 1990.

Dauben, Joseph Warren, *Georg Cantor: His Mathematics and Philosophy of the Infinite*. Princeton, New Jersey: Princeton University Press, 1979.

Dedekind, Richard, *Essays on the Theory of Numbers*. La Salle, Illinois: Open Court Publishing Company, 1948.

Deloche, Gérard, and Seron, Xavier, eds., *Mathematical Disabilities: A Cognitive Neuropsychological Perspective*. London: Lawrence Erlbaum Associates, 1987.

de Villiers, Peter A., and de Villiers, Jill G., *Early Language*. Cambridge: Harvard University Press, 1979.

Euclid, *Elements*. New York: Dover Publications, 1956.

Fiedel, Stuart J., *Prehistory of the Americas*. New York: Cambridge University Press, 1987.

Flegg, Graham, *Numbers: Their History and Meaning*. New York: Schocken Books, 1983.

Flegg, Graham, *Numbers Through the Ages*. London: MacMillan Education LTD, 1989.

Freeman, Kathleen, *Ancilla to the Pre-Socratic Philosophers*. Cambridge: Harvard University Press, 1966.

Gillings, Richard, *Mathematics in the Time of the Pharoahs*. New York: Dover Publications, 1972.

Gless, Paul, *The Human Brain*. New York: Cambridge University Press, 1988.

Gleick, James, *Chaos: Making a New Science*. New York: Viking Penguin, 1987.

Griffin, Donald R., *Animal Thinking*. Cambridge: Harvard University Press, 1984.

Hadamard, Jacques, *The Psychology of Invention in the Mathematical Field*. New York: Dover Publications, 1945.

Harth, Erich, *Windows on the Mind: Reflections on the Physical Basis of Consciousness*. New York: William Morrow and Company, 1982.

Heath, Sir Thomas, *A History of Greek Mathematics*. London: Oxford University Press, 1921.

Hobbes, Thomas, *Leviathan: Part I and II*. New York: Bobbs-Merrill Company, 1958.

Hollingdale, Stuart, *Makers of Mathematics*. London: Penguin Books, 1989.

Huntley, H. E., *The Divine Proportion*. New York: Dover Publications, 1970.

Ingham, A. E., *The Distribution of Prime Numbers*. New York: Cambridge University Press, 1990.

James, Glenn, and James, Robert, eds., *Mathematics Dictionary*. New York: D. Van Nostrand Company, 1959.

Jaynes, Julian, *The Origin of Consciousness in the Breakdown of the Bicameral Mind*. Boston: Houghton-Mifflin Company, 1976.

Kalmus, H., "Animals as Mathematicians," *Nature 202* (June 20, 1964.)

Kamke, E., *Theory of Sets*. New York: Dover Publications, 1950.

King, Jerry P., *The Art of Mathematics*. New York: Plenum Press, 1992.

Kline, Morris, *Mathematics: A Cultural Approach*. Reading, MA: Addison-Wesley Publishing Company, 1962.

Kline, Morris, *Mathematical Thought from Ancient to Modern Times, Vol. 1*. New York: Oxford University Press, 1972.

Knopp, Konrad, *Infinite Sequences and Series*. New York: Dover Publications, 1956.

Kröner, Stephan, *The Philosophy of Mathematics*. New York: Dover Publications, 1968.

Kosslyn, Stephen M., and Koenig, Oliver, *Wet Mind: The New Cognitive Neuroscience*. New York: The Free Press, 1992.

Lambert, David, *The Field Guide to Early Man*. New York: Facts on File Publications, 1987.

Leakey, Richard E., *Origins*, New York: E. P. Dutton, 1977.

Lemonick, Michael D., "*Fini* to Fermat's Last Theorem," *Time*, 5 July 1993, p. 47.

Lewin, Roger, *Bones of Contention: Controversies in the Search for Human Origins*. New York: Simon and Schuster, 1987.

Lumsden, Charles J., and Wilson, Edward O., *Promethean Fire: Reflections on the Origin of the Mind*. Cambridge: Harvard University Press, 1983.

MacLean, Paul D., *The Triune Brain in Evolution*. New York: Plenum Press, 1990.

McLeish, John, *Number*. New York: Fawcett Columbine, 1991.

Menninger, Karl, *Number Words and Number Symbols: A Cultural History of Numbers*. New York: Dover Publications, 1969.

Moffatt, Michael, *The Ages of Mathematics, Vol. I, The Origins*. New York: Doubleday & Company, 1977.

Muir, Jane, *Of Men and Numbers*, New York: Dodd, Mead, & Company, 1961.

Newman, James, ed., *The World of Mathematics*. New York: Simon and Schuster, 1956.

Newsom, Carroll, *Mathematical Discourses*. Englewood Cliffs, NJ: Prentice-Hall, 1964.

Niven, Ivan, *Numbers: Rational and Irrational*. Washington, D.C.: The Mathematical Association of America, 1961.

Ottoson, David, *Duality and Unity of the Brain*. New York: Plenum Press, 1987.

Pappas, Theoni, *The Joy of Mathematics*. San Carlos, CA: World Wide Publishing/Tetra, 1989.

Peter, Rozsa, *Playing with Infinity*. New York: Dover Publications, 1961.

Plato, *The Dialogues of Plato*, trans. B. Jowett. New York: Random House, 1937.

Preston, Richard, "Profiles, The Mountains of Pi," *The New Yorker* (March 2, 1992).

Resnikoff, H. L., and Wells, R. O., Jr., *Mathematics in Civilization*. New York: Dover Publications, 1984.

Robins, Gay, and Shute, Charles, *The Rhind Mathematical Papyrus*. New York: Dover Publications, 1987.

Rucker, Rudy, *Infinity and the Mind*. New York: Bantam Books, 1982.

Russell, Bertrand, *The Problems of Philosophy*. London: Oxford University Press, 1959.

Sacks, Oliver, *The Man Who Mistook His Wife for a Hat*. New York: Harper Perennial, 1985.

Sagan, Carl, *Mind in the Waters*, ed. Joan McIntyre. New York: Charles Scribner's Sons, 1974.

Schmandt-Besserat, Denise, *Before Writing, Vol. I: From Counting to Cuneiform*. Austin, TX: University of Texas Press, 1992.

Schmandt-Besserat, Denise, "The Earliest Precursor of Writing," *Scientific American* (June 1978).

Smith, David Eugene, *History of Mathematics, Vol. 1*. New York: Dover Publications, 1951.

Smith, Steven B., *The Great Mental Calculators*. New York: Columbia University Press, 1983.

Smith, T. V., and Grene, Marjorie, eds., *Philosophers Speak for Themselves: From Descartes to Locke*. Chicago: University of Chicago Press, 1957.

Soustelle, Jacques, *Mexico*. New York: World Publishing Company, 1967.

Stein, Sherman K., *Mathematics: The Man-Made Universe*. San Francisco: W. H. Freeman and Company, 1963.

Treffert, Darold A., *Extraordinary People: Understanding Idiots Savants*. New York: Harper & Row Publishers, 1989.

Werkmeister, W. H., *A Philosophy of Science*. Lincoln, NE: University of Nebraska Press, 1940.

Whitehead, Alfred North, and Russell, Bertrand, *Principia Mathematica*. Cambridge, England: Cambridge University Press, 1950.

Winick, Charles, *Dictionary of Anthropology*. Totowa, NJ: Littlefield, Adams, and Company, 1970.

Woodruff, Guy, and Premack, David, "Primitive Mathematical Concepts in the Chimpanzee: Proportionality and Numerosity," *Nature 293* (October 15, 1981).

Zaslavsky, Claudia, *Africa Counts*. New York: Lawrence Hill Books, 1973.

Zippin, Leo, *Uses of Infinity*. Washington, D.C.: The Mathematical Association of America, 1962.

Index